The Royal Small Arms Factory

Enfield

&

Its Workers

1 Late in the year 1938 Bren gun parts were being delivered for assembly at Enfield in large quantities from BSA Guns Ltd

The Royal Small Arms Factory
Enfield

&
Its Workers

David Pam

To Valerie Carter with my warmest thanks for her help with all my books.

ISBN 0 9532271 0 3

Published by the author and available from:
Jubille Hall
2 Parsonage Lane
Enfield
Middlesex EN2 0AJ

Designed and produced by John Robey, Mayfield Books, Mayfield, Ashbourne, Derbyshire

Printed in Spain by
Keslan Servicios Graficos

Contents

Illustrations, Maps, Plans & Elevations

Note on Sources

Parliamentary Papers (PP) and Command Papers (C, Cd, Cmnd) are available in the British Library, State Papers Department. Documents from the War Office (WO), Ministry of Supply (SUPP) and Ministry of Munitions (MUN) are available in the Public Record Office.

Acknowledgements

I am beholden to a number of former employees at the factory for comments, information and advice. Many of them, members of the Royal Small Arms Factory Apprentices Association, were put in touch with the author through the good offices of Ray Tuthill to whom I owe very special thanks. Among those who have helped in this way I must name particularly: Ossie Edwards, Stan Carroll, Bert Hunt, Roy Burgess, Charlie Brown, Ron Everton and Denis Hardy. Others who have provided help include Geoffrey Gillam, Patricia Lovell and Herbert Woodend, Keeper of the Pattern Room.

British Aerospace (Royal Ordnance) at Westcott has provided many of the illustrations, my thanks to Ginny Willcox and Richard Culmer. Other illustrations have been borrowed from L.B.Enfield Libraries with the indefatigable help of Kate Godfrey and Graham Dalling. The Enfield Preservation Society has supplied me with further illustrations and help, I have particularly to thank Stan Smith and Irene Smith, the late and irreplaceable secretary of that society, on whom I depended so much in the publication of my *History of Enfield*. I am moreover grateful to Stephen Sellick and to Barbara Bassett who not only provided me with many photographs but carefully explained the scenes depicted.

Much of the research was done at the Public Record Office and the British Library (State Papers and Newspapers depts) where the staff have been both courteous and helpful. Illustrations from the Public Record Office can be identified by the document references provided in the list of illustrations.

I have frequently consulted the very learned work on the RSAF by Tim Putnam and Dan Weinbren (published by Middlesex University) and have been guided by their greater understanding of industrial processes.

Valerie Carter has read my work and corrected my mistakes where possible.

Christopher Simons has somehow contrived to put it all on to a small computer disk which it seems to me, must be a far more difficult task than the mere writing of this book.

Introduction

This history, from the beginning of the RSAF in the years following the defeat of Napoleon until its closure in 1988, is concerned more with the men who worked at Enfield Lock than with the weapons they made. The conditions, in those early years, were determined by a location so remote that the essentials: housing, education and religion, had to be provided on the site. The need to retain a nucleus of skilled workmen in peacetime, as an insurance in time of war, directed policy over the first forty years and was later reaffirmed on many occasions. At the same time the Government felt it necessary to foster a private small-arms industry by the allocation thereto of the greater proportion of a limited number of orders for military weapons, otherwise the country would soon have found itself without the capacity to produce them except within the government factory, but this it did not wish greatly to enlarge in times of peace. The private manufacturers however constantly colluded in fixing prices. Even worse, they placed the nation in peril, both during the Napoleonic and the Crimean wars, by their repeated failures to complete government orders on time.

At the 1851 Exhibition six American army rifles, manufactured by Robbins and Lawrence with interchangeable parts, had been displayed, they created widespread interest. The possibility of imitating the Americans and at the same time remedying the supply situation was considered by the Royal Small Arms Commission under the chairmanship of James Nasmyth set up in 1853. At that time the parts for small-arms in this country were made by workmen in attics and back rooms to be collected up and assembled, each part having to be laboriously adapted to fit.

A report by Whitworth and Wallis in 1854 on interchangeability in American industry, following the New York Exhibition of 1852, led to a visit by John Anderson, Col Burn and Capt Warlow. They travelled widely throughout the United States observing closely the facility with which innovations and methods of mass production had been introduced in a wide variety of manufactures. They predicted that US industry would, before long, overtake European industry, and even that of England. The Government, impressed by their enthusiasm for American methods, accepted the recommendation they had made to purchase machinery in that country to equip a new factory for the production of small-arms using interchangeable parts.

The gun trade now conspired to ensure that the new factory would not be built at Woolwich, for there its competition would inevitably have proved damaging to commercial interests. Thus they contrived that it would be built at Enfield Lock. The site there, isolated, flat and swampy, would, they thought, be unlikely to attract and retain skilled artisans. Many of the new workers were forced to take poor accommodation at Enfield Wash, the nearest settlement along the main road. It was described by a correspondent in 1857 as 'Nothing but open privies and cesspools, and the sewage from the houses runs from the cesspools into the wells'. There was no supply of pure water. 'The dwellings along the High Road from the Sun and Woolpack to St James' Church, including those in the alleys and yards leading off the main road, are crowded with humanity, due', he went on, 'to the great addition lately made to the Ordnance Factory .... The stench from the open ditches and from the filth on the sides of the highway, caused by the long accumulation of human and horse manure was insufferable in hot weather.' Despite its disadvantages the Enfield factory was an immediate triumph and by 1860 it was manufacturing five times as many military weapons as all other sources combined. Representatives of aspiring states were sent to Enfield, desperate to acquire, by fair means or foul, the secrets of mass production.

The import of American machinery capable of producing interchangeable parts led, within a very few years, to the manufacture of similar machines in this country. Thus it was that when, in 1861, stimulated by the demand created by the American Civil War, sixteen Birmingham gun makers built a new factory at Small Heath, they were able to order the necessary machinery, to the value of £6,101, from Greenwood and Batley. The same firm, two years later, exported machinery worth £30,319 to the United States. Thus the government enterprise at Enfield had already done much to stimulate, not only the small-arms industry, but also the British machine-tool industry.

The factory continued to play an important role in improving the design of machinery. Thoughout the years up to the First World War, repeated changes in the design of small-arms caused difficulties both for management and men. Production slowed down or stopped with every change of design and workers lost wages. Yet each change demanded capital investment in new machinery and created, on completion, a surge in demand. To keep costs down, new production methods were constantly sought. New machinery and improvements to existing machinery were often designed, constructed and tested within the factory. If they worked and more were required a contract would be placed with a specialist machine-tool maker. Thus despite difficulties in rewarding invention in a government establishment, and the expense, in this country, which precluded workmen securing patents, the management succeeded in keeping the factory in the forefront of developments in the small-arms industry.

A tradition had existed since 1854 that the principal infantry weapon for the British Army should be designed and made at Enfield. There had been the Enfield rifle, which served the country so well in the Crimean war. It was superseded in 1864 by the Snider, the Enfield rifle adapted for breech-loading. This was replaced by the Martini Henry, a newly designed breech-loader, in 1871, and in 1889 by the Lee Metford. By 1895 came the Lee Enfield, followed in 1902 by the Short Magazine Lee Enfield (SMLE), the most famous of all the Enfield weapons. It was to last for fifty years and was to see the country through two world wars. Over two million of them were produced at Enfield during the Great War 1914-18, when at times the factory employed more than ten thousand workers. During the Second World War hundreds of skilled artisans from the RSAF were sent out like missionaries to carry their expertise into new factories built to maintain the supply of small-arms to our forces. Those of us who served in the Army (or in the Home Guard) were constantly reminded of the RSAF by the omnipresence of the Bren gun and the Sten gun whose names incorporated the 'EN' of Enfield. The SMLE was not finally replaced until the 1960s when the FN rifle was introduced; this gave way, in 1985, to the SA80, the Enfield Weapons System (EWS).

Stories of privatization were heard in 1980; it was the beginning of the end of the story. By 1986 there were rumours of impending closure. The factory finally closed in 1988; in the words of T. S. Eliot, 'not with a bang but a whimper'.

Chapter 1
Beginning at Enfield Lock

In the seventeenth century the Government purchased whatever arms it required from contractors either in Birmingham or London, but it was essential that the Board of Ordnance should hold adequate stores of weapons and parts, for otherwise the State would certainly find itself at the mercy of the contractors in any emergency. The building of a Grand Storehouse in the Wardrobe Garden at the Tower of London, to contain also a viewing-room and workshops for furbishers, was begun in 1663, and a new proof-house on the Tower Wharf was added in 1682. The ordering of complete weapons was abandoned in the early eighteenth century, only orders for component parts were to be issued, the barrels and locks from Birmingham and the woodwork from London. All the parts when delivered were stored at the Tower from where the setting-up was carried out by gunmakers in the neighbourhood. The system was under the management of the Principal Storekeeper of the Small Gun Office, above him in the hierarchy was the Master Furbisher, who was often called upon to make modifications to the weaponry, either suggested by himself or by others. The Small Gun Office had greatly expanded by 1750 and the group of workshops built within the Tower had assumed the proportions of a small factory. At times a hundred gunmakers were employed there in cleaning, repairing and on the assembly of parts to make new weapons; some were paid by the day others were on piecework.

Because muskets made for the Army had to be of a standard pattern, and because they had to be reliable and straightforward to repair, the Small Gun Office found it necessary to retain control of manufacture by means of inspection. The stringency of the examination of gun barrels by the Ordnance at the Tower gave rise to resentment among the Birmingham contractors. Returns in 1764 show that one barrel in ten was rejected at view and a further percentage at proof. The contractor had thus to suffer the loss of rejected barrels as well as the cost of their transport from Birmingham. From 1777 the viewers were accommodated at Birmingham but their inspection became increasingly stringent and the percentage of failures rose. The contractors, growing ever more incensed, retaliated by giving government orders an ever lower priority. Thus the supply of parts from private industry diminished and in 1779, during the war with the American colonies, the British government was obliged to send an agent, John Colgate, to Liège to superintend the manufacture of 40,000 muskets there.

By 1 February 1793 when the war with revolutionary France began, the supply of arms was still too unreliable for national security.

> With all the progress made in the production of arms, the Board of Ordnance remains at the mercy of the contractor with his erratic deliveries, his quarrels with his workmen and his manipulation of prices,

wrote the Secretary to the Board on 25 July 1794.

> The Board is perfectly convinced that the only method that can be taken to prevent, in the future, the present complaint, is to have a manufactory of small-arms upon the establishment of the Ordnance which, besides producing a supply of arms in case of emergency, would become a check upon the proceedings of the gunmakers and prevent combinations among them against the Government.

Such an establishment however would take years to prepare. Colonel Miller, in 1795, had to be sent to Liège and Hamburg desperately seeking to purchase muskets. The situation remained perilous and when calls began to multiply for arming volunteers and local militia, in case of invasion, Colonel Miller was again dispatched to Hamburg in 1800, to procure more arms. Lord Chatham acknowledged publicly in 1802, when he was Master General, that the art of making military small-arms had been lost in this country. The Government was forced to continue to purchase from the Continent and by 1804 about 300,000 stands of very clumsy and inferior muskets had been imported.

The Small Gun Department, on 1 April 1804, came under the control of the newly appointed Inspector of Small-Arms, Lt-Colonel Miller. An assistant inspector, Mr Alsop and many viewers and labourers were taken on, both at Birmingham and at the Tower. The most important duty of the new department was the inspection and improvement of arms furnished to the Ordnance. It assumed control of the workshops at the Tower where, in 1805, it took on a staff of rough-stockers, setters-up and makers of small materials, with John Noble, an experienced viewer, as superintendent.

The difficulty of securing an adequate supply from Birmingham however did not diminish and at last Parliament was persuaded, in 1805/6, to vote money for the purchase of the Lewisham Mills. It had been a plate mill for the supply of rough plate to the Royal Armourers at Greenwich in the sixteenth and seventeenth centuries, up to 1685. Armour by that time was becoming obsolete and the mill had been let to tenants. In 1729 it had been making tools for building military roads in Scotland and subsequently swords and hafted weapons had been made there. Following the take-over by the Ordnance, Thomas Fullerd, the Clerkenwell barrel-maker, became superintendent at a salary of £150 a year. Despite the requirements of war, the Lewisham factory could not be made ready for its new function until 1808, in the meantime the Ordnance used Fullerd's factory in Allen Street Clerkenwell, off Goswell Street. It was not Fullerd however who eventually took charge at Lewisham, but John Colgate at £188 a year. The conversion had cost £66,000 but the factory was not a success. There had been doubts in the summer of 1807 concerning the sufficiency of water-power at Lewisham, especially in view of a proposal by Captain Mulcaster and John Colgate to expand the capacity for production by the addition of a second mill. John Rennie, the well-known engineer, was asked to investigate. He found that the flow of water from the River Ravensbourne would not be sufficient for even one mill in the drier months of the year. It was the Board's opinion that a steam-engine would have to be provided and Captain Mulcaster, urging that the factory should commence work before the end of the year, for by that time the Army had become involved in the Peninsular War, required a decision on the power that would be required to keep the manufactory in full employ. Mr Lloyd, of Lloyd and Ostell, was asked to construct a steam-engine to produce 36hp, and to submit an estimate. The engine, which cost £2,400, was in operation by the end of March 1809 and although it 'appeared to answer fully', by reason of high running costs (£2,000 a year) it failed to render the factory efficient.

Meanwhile Birmingham production, faced by the threat of government competition, increased dramatically. However in order to achieve the increase, the system of strict view and inspection had been relaxed, for it had not been found possible to procure enough artisans capable of working to the high standard which had previously been insisted on for military weapons. Ordnance now required the Birmingham masters to set up arms complete, whereas for some time they had been confined to making parts. Their exertions at least produced the quantity of arms demanded and, by the end of the war, production was to reach a thousand muskets a day, but the regular army pattern rifle had had to be abandoned for the more easily obtained India pattern of poor quality.[1]

A parliamentary committee in 1811, appointed to consider the output at Lewisham, expressed strong criticism concerning the spending of £66,000 which had resulted in the production of no more than 25,000 barrels and 18,000 locks a year, against more than ten times those quantities made in Birmingham. It was suggested that the Board of Ordnance should inquire whether it was expedient to continue to manufacture at Lewisham, indeed whether a government department should engage in arms manufacture at all. But the Board, remaining convinced of the necessity for such a check on private manufacturers, required Major Mulcaster in February 1811, to estimate the expense of moving the works to the neighbourhood of Cheshunt where it could be powered entirely by water. Mulcaster in his reply pointed out that the water available north of Cheshunt was essential to the operation of the Waltham Abbey Powder Mill, which had been taken over by the Board of Ordnance in 1795. He suggested, in preference, a site to the south, near the fishery which the Ordnance had purchased from Mr Sotheby. From there, communication to the Tower would be less by four miles, far more water would be available to power the mill, and there was land for future expansion. He pro-

2, 3 The print shows the orginal Swan and Pike which later became the superintendent's house, the photograph shows the second Swan and Pike, formerly the lock-keeper's cottage

duced an estimate to include water-mills and grinding sheds, a building to accommodate 24 barrel forges, a building to house the loop and sight forges and the women smoothers' shop. It amounted in total to £28,344. There was concern that there should be no wastage of water by river transport. To this end legislation was drafted that lock-keepers be appointed at each of the locks from Kings Weir to Enfield lock and that accommodation for them be built so that the locks should be manned day and night, also that the locks be converted from pound locks to chamber locks to prevent waste.[2]

John Rennie was asked, in April 1811, to report giving his opinion on the power of the water and the eligibility of the situation. The site was immediately below the conflux of the Small Lea and the Old Lea and thus it had the use of water from both streams with a fall of seven feet nine inches; this Rennie considered likely to produce ninety horsepower. There was access by water and, being close to the Royal Powder Mills, one set of barges could serve both factories. Negotiations were begun at once with Charles Smith, who was the freeholder. Smith's land was leased to Augustin King George, the miller at Ponders End, and he in turn had sublet much of the property to Thomas Wace. This included the house where Wace lived, which had formerly been a public house called the Swan and Pike. Wace demanded £600 for the remainder of his term. It was urged that the Board should settle the matter quickly before Wace realized the potential profitability of acquiring a new licence for the Swan and Pike, for 'gun barrel makers', it was observed, 'will have a drink whether it is to be procured from far distant or near at hand', and there was not another public house within a mile of the site. It was thought unlikely that there would be any difficulty in securing a new licence, and it was felt that the Board might itself wish to let the building as a public house, for not only would it fetch a very good rent, but the landlord would be under the control of the Board and could readily be displaced for any conduct of which the Board might disapprove.[3]

An agreement was reached by 11 January 1812 on a price of £3,156 out of which Smith was to compensate his subtenant. Minor difficulties remained even then, for a footpath to Sewardstone passed through the site crossing the Old River Lea by a footbridge where a penny toll was exacted. It would have to be closed. There was moreover a fishery let by Charles Smith to one William Clarke for £10 a year. He sublet it to various subscribers who would not be welcome when the new factory was built. Clarke would have to give up that part of the river. Captain John By was commanded to inspect the site. He had taken over, from Major Mulcaster, command of the Royal Engineers stationed at the Royal Powder Mills Waltham Abbey. In his later years he became famous for his engineering works in Canada where he gave his name to Bytown, subsequently renamed Ottawa. John By's first visit to Enfield Lock was made early in June 1812. At that time his intention, to take an accurate survey of the ground, was frustrated by another subtenant, Griggs, who refused to allow him to walk over the

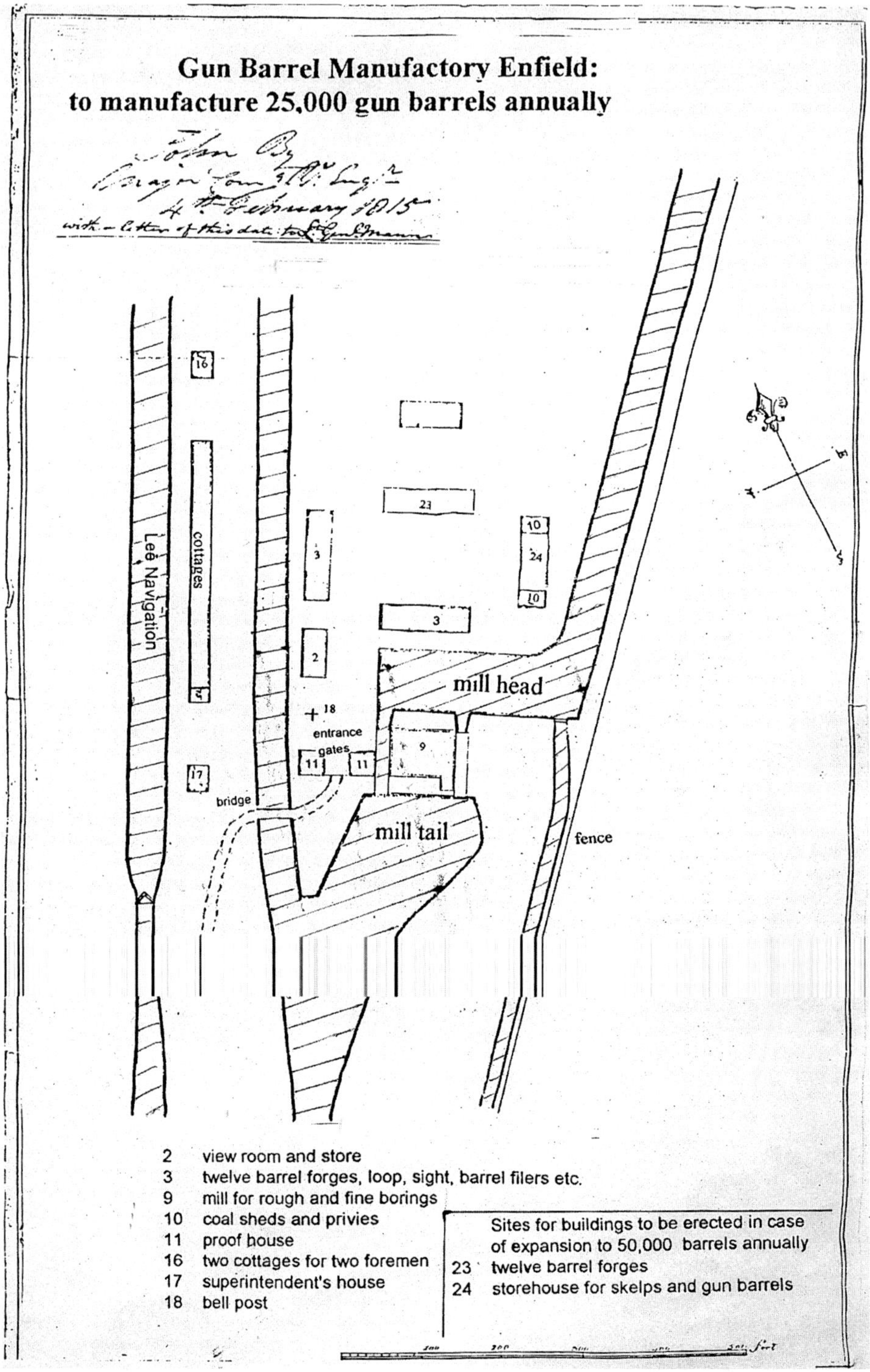

4 *John By's plan 4th February 1815*

land because the meadow had yet to be mown. He was nevertheless able to make a sketch of the ground and he thought this would be good enough to formulate a plan to indicate the positions of the proposed buildings. The Board, having acquired this desolate stretch of marsh, had to station a man there to guard it against the depredations of 'idle or ill-disposed persons'. Wharfing would have to be done before the frosts or the floods would wash away the land. Thus 17 acres was acquired in the parishes of Sewardstone and Waltham Holy Cross and marked out by boundary stones bearing the King's broad arrow. The pound upon the island site belonging to the Sewardstone silk mill was purchased. Also land abutting the Lee Navigation, this in order to avoid disputes with the Navigation surveyor, 'a very petulant and overbearing upstart'. Traffic through Enfield lock was not overwhelming, three or four boats a day through the week, eight to ten on a Sunday.

Captain By was also asked to consider, with advice from Alsop the Inspector of Small Arms, what machinery could be removed from Lewisham to Enfield and adapted there to the use of water-power. He was further required to investigate and decide what additional machinery would have to be purchased. He was then to confer with Messrs Lloyd and Ostell, machine-tool makers of Gravel Lane

Plan and Elevation of the Storehouse for Skelps and two Coalsheds and Privies

Blackfriars Road, as quickly as possible, for the preparation of new machinery was likely to take some time.[4]

Lloyd and Ostell, by June 1812, had designed a new lathe for turning barrels and it was enthusiastically recommended by Captain By. 'It answers in every respect', he wrote, 'the barrels turned by it being much better finished than those finished at Lewisham by grinding and filing'. Barrels produced on the new lathe, he thought, would require far less labour, and the cost would be lower. Lloyd's new lathe was lighter and less expensive than those used at Lewisham for grinding the barrels. 'These lathes will last for centuries', said By, 'and any man may learn to turn barrels in a few lessons'. He calculated that each barrel would require 30 minutes to be fixed into the lathe, turned and taken out, so that 20 could be produced in a ten-hour day or, over 313 days, 75,120 barrels. It was so easy, he asserted, that children could be taught. The estimated cost of this new machinery, and the installation of machinery from Lewisham at Enfield was £40,735, to which would have to be added the cost of removal. Rennie proposed the excavation of a navigable canal one-third of a mile long to protect the water-wheels from flood, also to provide access for barges.[5]

Employed at Lewisham in the barrel-making branch were 36 welders, 18 rough-

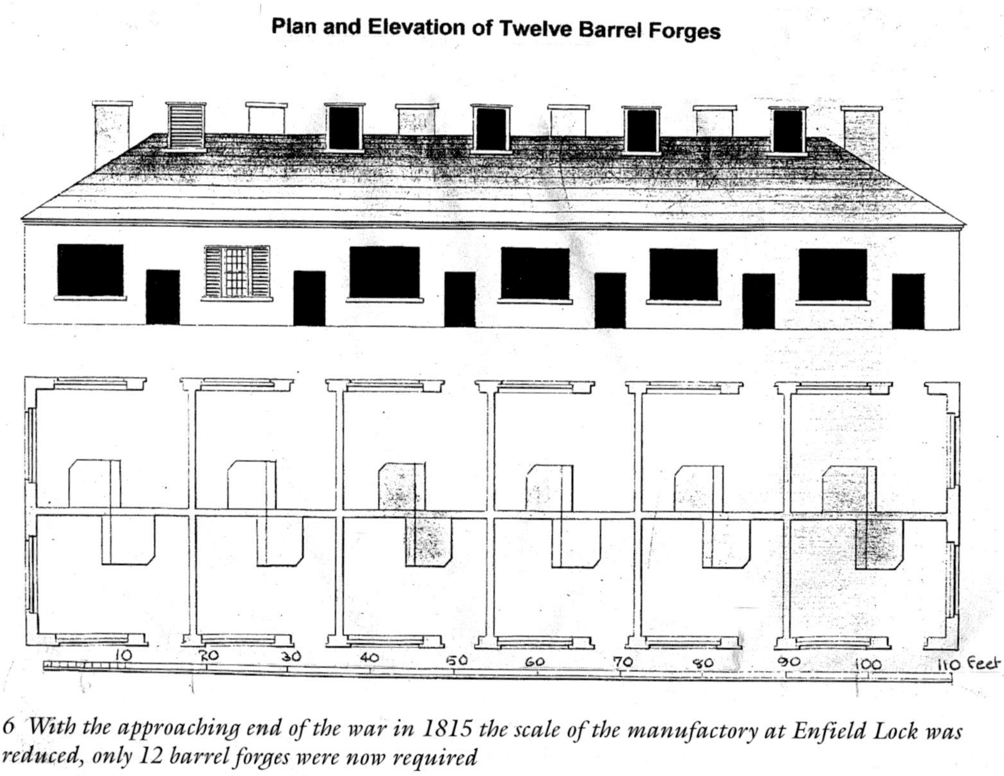

6 With the approaching end of the war in 1815 the scale of the manufactory at Enfield Lock was reduced, only 12 barrel forges were now required

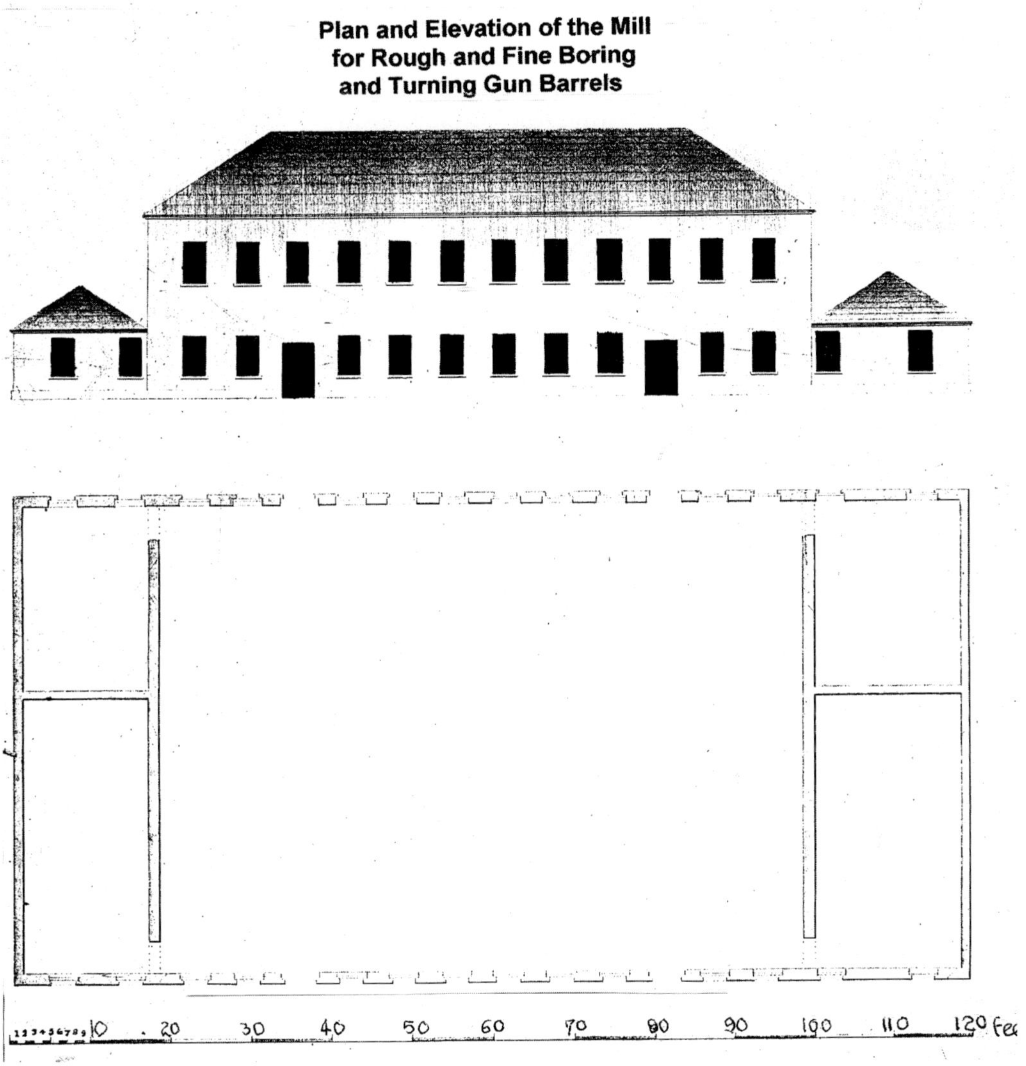

7 The mill for rough and fine-borers was completed by November 1815 except for the parts to come from Lewisham

borers, 12 smooth-borers, 12 grinders, 30 filers and 20 women and boys. Alsop was asked to make enquiries locally whether the workmen would be able to procure lodgings in Enfield, or whether it would be necessary to build cottages on the site. A cursory glance at the area might have told him that there was no possibility of rented accommodation. The nearest settlement was at Enfield Wash but the road leading there was often under water and impassible. It would have to be raised and repaired but it was considered, with some justification, that the parish would be unwilling to pay for the work. So it was that Captain By was asked (19 August 1812) to prepare plans and estimates for cottages and for a superintendent's house.[6]

Rennie, following consultation with Captain By, had prepared two factory plans by July 1813, but the Board considered them elaborate and expensive. At this stage it was merely planned to build a barrel manufactory at Enfield equal to that at Lewisham. It was the Board's opinion that one mill with twelve lathes would be sufficient to produce the 50,000 barrels a year which it calculated would be required. Nevertheless Rennie's first plan, which included the canal, was accepted regarding the position of the mills and the construction of the mill head.

In February 1814 Captain By was ordered to proceed with the move from Lewisham as quickly as possible. He was to prepare further plans showing the watercourses, the workshops and buildings and he was to draw up an estimate of the total cost of the buildings. In March Lloyd and Ostell submitted their estimate for new machinery, and for the transfer of machinery from Lewisham. The firm was now required to provide two water-wheels each 18 feet in diameter but increased from 14 to 18 feet in width so that two sets of machinery could be operated separately. This change of plan was not to the liking of Lloyd and Ostell for one water-wheel had already been constructed, 14 feet wide, in accordance with earlier instructions, but John By pointed out that the 14-foot wheel would provide sufficient power for rough boring. Captain By's initial estimate was for £98,357 and this included 81 cottages. His plans were ready by the end of March and were passed for examination to Captain Dundas, the Assistant Inspector of Small Arms, and to Colgate the superintendant at Lewisham. The cottages were to be erected on the site, but this proved difficult for solid foundations were not possible, the ground being soft, consisting of peat and marl; solid ground could only be reached at a great depth. William Atkinson, the architect to the Board, recommended planking instead of foundations, but this was to matter less for the proposed cottages were soon to be reduced from two storeys to one. It was also considered necessary to raise the ground two feet front and back to keep the road and the floors of the cottages dry in times of flood. The war by now was drawing to an end. Peace negotiations through February 1814 proved unsuccessful but the Board was already considering modifications to its plans at Enfield Lock.[7]

It was decided in November that the buildings were to provide for the manufacture of only 25,000 barrels a year, though the site was to be laid out in such a manner that production there could readily be increased to 50,000. Major By was ordered to revise his plan of 4 April and to transmit new plans and estimates. The reduction of scale eliminated the need for a planned storehouse for skelps and gun barrels, 12 of the barrel forges were no longer required nor a superintendent's house, but sites were allocated for these, in case of any emergency, in By's plan of February 1815. The Allies entered Paris on 30 March and, on 11 April, Napoleon abdicated and received the princedom of Elba. His final defeat at Waterloo occurred on 18 June 1815.[8]

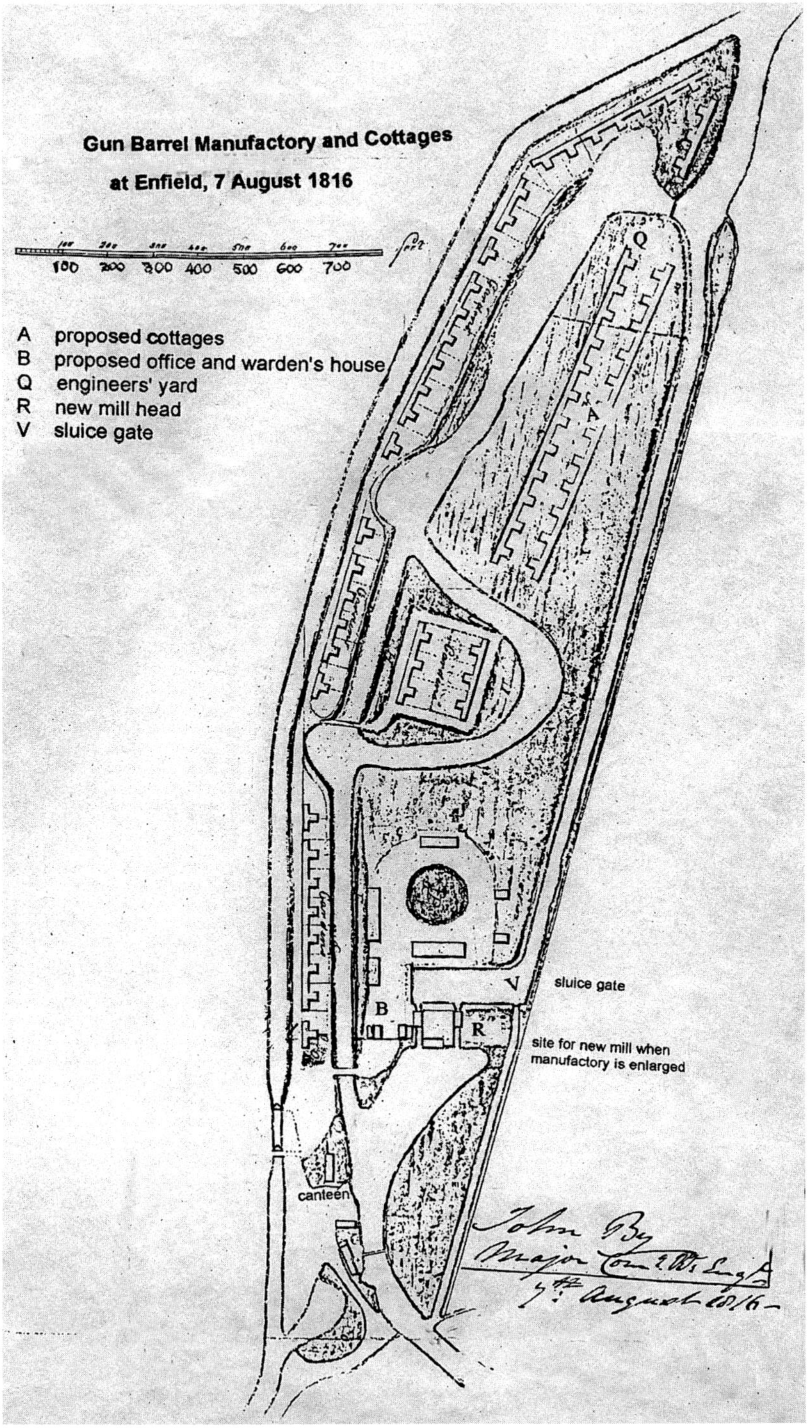

8 Major John By's plan August 1816

The end of the war saw a virtual cessation of the demand for small-arms. The work-force at Birmingham was halved, and the Inspector of Small-Arms returned to London.

The consequence of the high wartime production at Birmingham was that by the time outstanding contracts were called in after the war, there were so many muskets in store that the manufacture of military arms had to be entirely suspended. Many of the best workmen were in consequence enticed away to America and elsewhere, others sank to the making of common 'Brummagen ware', called 'birding pieces', or 'trade' or 'slave' guns, sold as cheaply as 7s 6d each. Nevertheless building work continued at Enfield following the adoption of By's plan of February 1815 which showed savings amounting to £26,781 on By's original estimate.

By the beginning of November 1815 the view room, the black store and the shop for women smoothers were completed except for the interior fittings, and the building to accommodate twelve barrel forges was finished except for the four forges to be brought from Lewisham. The building for the forges of hoops and sights, and the shop for the barrel-filers, were ready apart from the benches. The mill for rough and fine-borers was complete except for the parts to come from Lewisham. The mill-head, watercourses and bridges were expected to be in working order before the middle of that month. Sixty cottages and four foremen's houses were almost ready for occupation by March 1816. The Board discharged 133 men, 12 women and 31 boys from the Tower and Lewisham on 21 March. They included barrel forgers and borers, lock-filers, forgers and polishers and gun-stockers, screwers-together and finishers.[9]

The following month Captain Dundas reported that the store of skelps, for making gun barrels at Lewisham, would be exhausted in the course of the ensuing week. It had been proposed that the welders should be moved to Enfield Lock at this juncture. He therefore suggested that two Waltham Abbey barges be sent to Deptford Creek Wharf on Wednesday evening, 24 April, to convey the men their families and furniture, and such of their tools as had not already been removed, to Enfield Lock. He proposed to the Board that since the foremen, the welders and their hammermen would lose money through the move, they should be paid three days average earnings on the day of removal; 7s 6d a day for the foremen, 7s a day for the welders and 5s for the hammermen and the same sum daily until work could be resumed in perhaps four or five days. He directed that Colgate should be present at Enfield to oversee operations, to allocate dwellings to workmen, and to set the men to work. He suggested that the men should be charged 2s a week rent for their cottages and that this should be deducted from pay.[10]

The first party from Lewisham was at work in Enfield by the middle of May. The second contingent moved at the end of that month with all the remaining barrel-making machinery; the barges had to make two journeys to Deptford Creek.

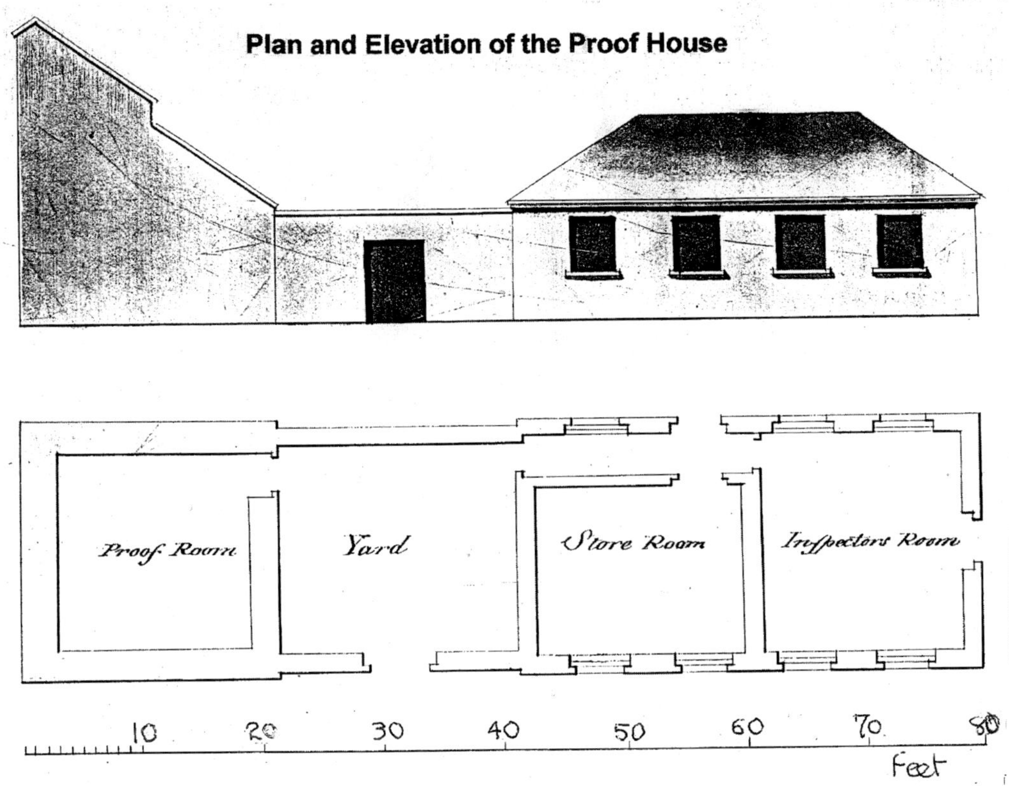

9 The proof-house at Lewisham was demolished in 1817 and re-erected at Enfield

Some of the cottages had scarcely been finished and the men were granted an allowance of coal to dry them out.

The firm Lloyd and Ostell was requested to send staff to Enfield to set up the machinery and to adapt it, where necessary, from steam power to water. The total spent on the manufactory at Enfield Lock, up to June 1816, amounted to £77,179. The proof-house at Lewisham was demolished in 1817 and re-erected at Enfield.[11] Thus the manufacture of barrels was moved but the transfer to Enfield was not finally completed until October 1818 when the Board decided that the two branches of small-arms manufacture at Enfield and Lewisham should be consolidated. Lewisham was therefore closed and the land, premises and the steam-engine were sold. Sixty-two men, including apprentices, were at that time employed at Lewisham of whom 22 had received the Board's promise of constant work in war or peace. These, with their apprentices, together with Benjamin Briscoe, the only man there capable of making tumbler locks, joined the work-force at Enfield, all others were discharged on 1 November with a month's pay and travelling allowances for themselves and their families. The women smoothers, for whom a

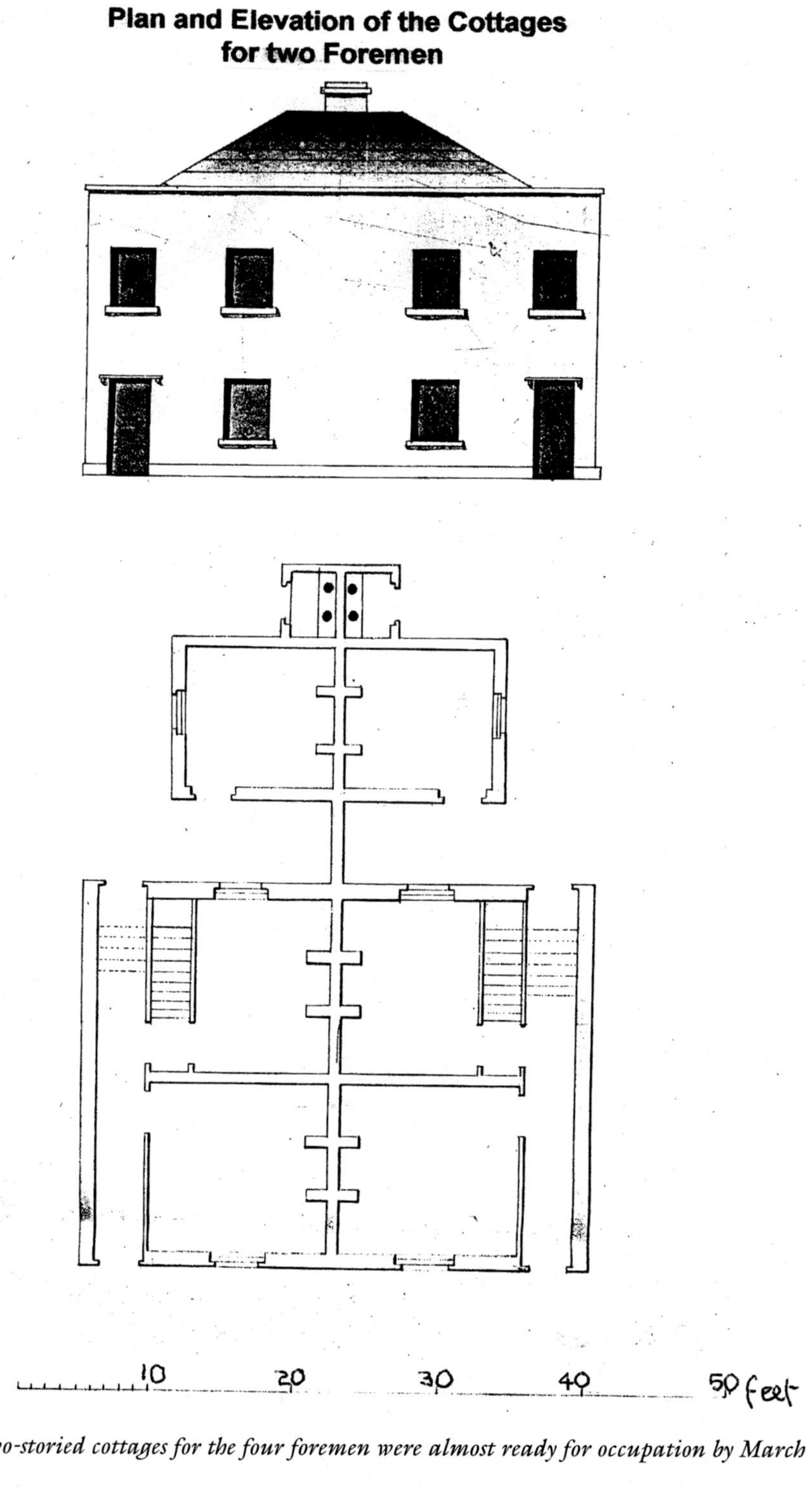

10 Two-storied cottages for the four foremen were almost ready for occupation by March 1816

workshop was being prepared in November 1815, were never taken on, for a policy of employing men only had been adopted in the factory. Three Waltham Abbey barges, one to be used as a coal barge, were sent to Deptford Creek to bring the men with their families, tools and furniture to Enfield Lock where 24 spare cottages and two foremen's houses were aired for their reception. The Board was of the opinion that no further reductions could be made at Enfield without impairing the nucleus of the manufactory.[12]

Notes to Chapter 1

1 Blackmore H. *British Military Firearms 1650-1850*, WO 44.682, PP 1854 xviii
2 PP 1810/1 iv 329, WO 55.753, WO 44.680
3 WO 44.680 5 Ap, 14 Jl 1811, Putnam T. and Weinbren D. *Short History of the RSAF* note 17
4 Robinson A. *Early days of the RSAF* p2, Putnam and Weinbren *op cit* p7, WO 44.680 14 Je 1812, WO 53.753
5 WO 44.680 6, 14 Je 1812
6 WO 53.753 29 My 1812, WO 44.680 29 My 1812
7 WO 44.680 23 Jl 1813, 9, 3 F 1814, WO 55.753 15 Mr 1815
8 WO 55.753 23 N 1814, WO 44.680 8 Mr, 4 F 1815
9 WO 44.680, Robinson *op cit* p7-8, WO 44.679 31 S 1815, WO 44.300
10 WO 44.679 17 Ap, 6 My 1816
11 WO 44.679 23 My 1816, WO 44.680
12 WO 44.300 23 Oct 1818

Chapter 2

George Lovell

George Lovell was appointed storekeeper at Enfield in 1816 and, in March 1817, the Board ordered the immediate removal of the whole of the Royal Engineers department, both workmen and workshops, and the transfer of all the premises to the storekeeper. Lovell was a remarkable man, widely read and able to translate German and French technical papers. From the beginning he set out to rearm the British forces with the best firearms in the world. Noble, the Enfield superintendent, held control over the workmen of the barrel branch at Enfield, while Colgate retained control of those transferred from Lewisham in 1818. It was an unsatisfactory arrangement but it persisted until Colgate retired in 1822. John Noble retired in 1824 and James Gunner, formerly the clerk at Lewisham, was then appointed superintendent under Lovell as storekeeper. Thus Lovell was left in undisputed control.[1]

Lovell's peculiar role was created by circumstances as well as by his own talents. The proper duties of a storekeeper were the care, ordering and payment for stores but, since the abolition of the posts of inspector and assistant inspector of small arms, all their duties had fallen to the storekeeper. In addition he hired artificers and fixed piece-work prices in the factory. It now had fifty employees. The management comprised the storekeeper at £350 a year, a clerk at £86 and the superintendent at £180. Each workshop was under the control of a foreman who also performed the duties of a viewer. There were three dwelling houses, for Lovell and his brother, the clerk, and for Gunner, 62 single-storied cottages and a canteen run by Tooley, described as the canteen keeper. At the behest of the Board of Health brick barrel drains were constructed in 1831 from the cottage privies into the river to remove 'the filth and night soil'. A few years later the factory became known as the Royal Armoury Mills. Only the best artificers were engaged, both that the works might set an example of perfection to private trade, and to ensure that such skilled men were retained in this country, for over forty first-class gunsmiths had emigrated to Russia and the United States following the end of the war. The most accurate machinery was now installed: boring and turning benches calculated to bore up to 50,000 musket barrels a year, rifling machines, barrel-turning lathes which not only produced barrels with unfailing accuracy but reduced the cost of production from 1s to 3d a barrel, also screw presses and stamps for preparing tools. In the sword department there were grindstones and polish-

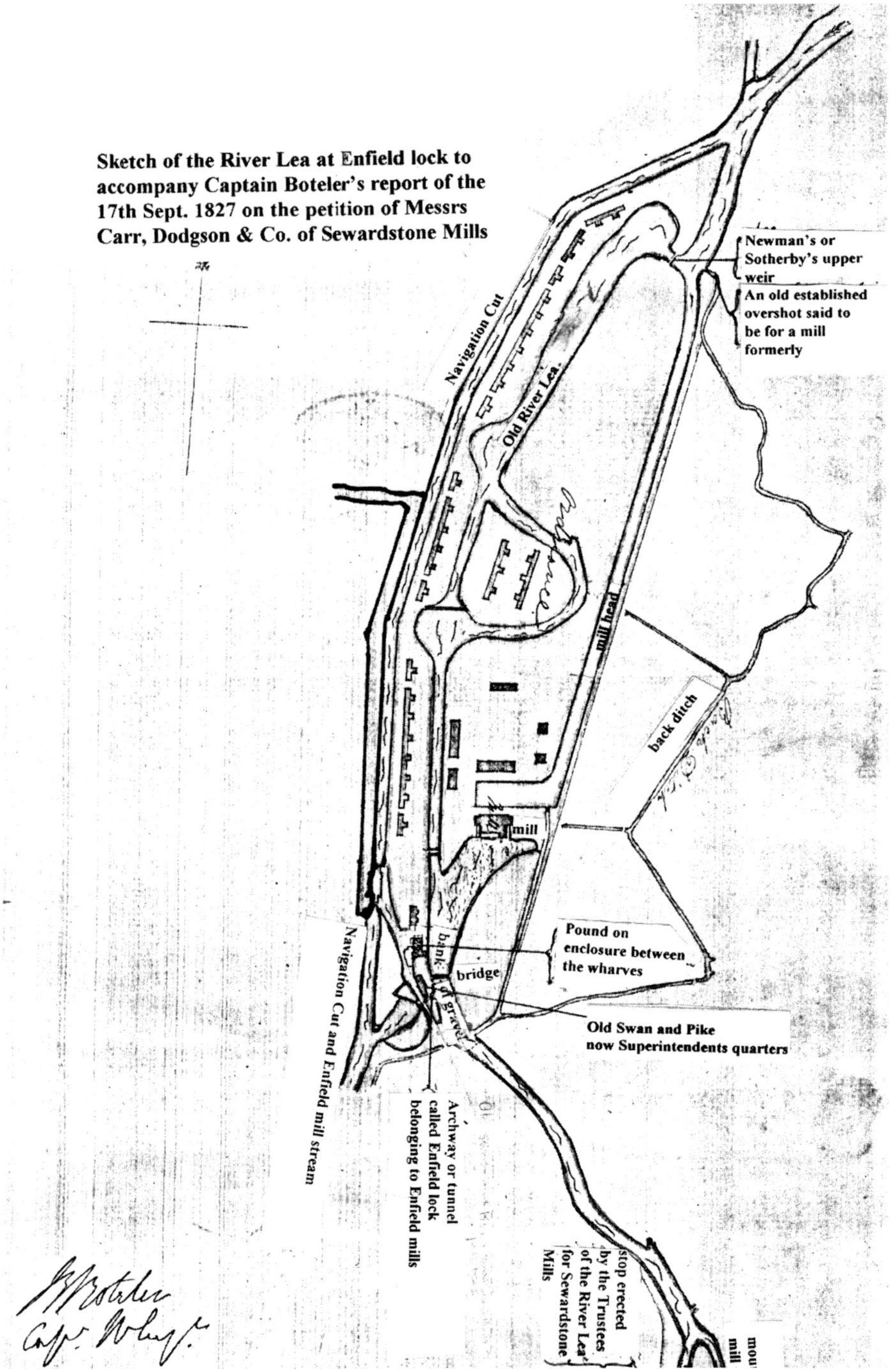

11 The factory was guarded night and day by a watchman. Workers could go in and out at leisure through the gateway, passing the sentry box (Robinson and Burnby Guns and Gunpowder in Enfield *p20)*

ing gear for swords, bayonets and other bright work. The mill, as described in 1823, consisted of two cast-iron undershot wheels, each providing 48 horsepower. Although there was capacity at Enfield to produce only 26,000 muskets a year, with a small expansion that number could easily be doubled. But the Board, in time of peace, opposed any extensions and when Lovell, in 1825, requested permission to increase the storage capacity for rough stocks, he was refused.[2]

Lovell pointed out that the land pattern musket in use at that time, the new Baker pattern, was a great improvement on the old English and Prussian arms, and he claimed that this improvement was due to the stimulus provided by the Royal Armoury Mills where 5,000 had been set up in 1823. The very existence of the government factory moreover had kept down the prices charged by private manufacturers and in doing so it had aroused bitter resentment, manifested recently in the hostility which greeted the setting-up of a sword branch at Enfield. When the contract for the new-pattern swords had been offered to the trade in Birmingham in 1823 there had been collusion between the contractors who had fixed a price of 31s 6d, but the sword branch at Enfield had been able to produce swords at 22s, and they were adjudged to be superior. The Board of Ordnance had been paying the trade 18s for rifling each barrel, but the work could be accurately executed at Enfield for only 8s 6d.[3]

Lovell would have liked to rival the production of the Birmingham combines. At the same time he set his artificers very high standards; they had to achieve an accuracy within one-thousandth of an inch. The factory continued the production of land pattern muskets (Baker pattern) through the year 1825 and at a lower unit cost than the trade, the factory claimed a profit of £1,405. It had serious limitations, however, for it was incapable of turning out a large order, and despite the fact that most of the parts could be made there, most orders for muskets went to Birmingham. The resulting situation became so desperate that in 1828 the factory found itself without work. Lovell, determined to keep going, sought whatever repairs and refurbishments he could get. The men were employed on the repair and conversion of old arms; five hundred hand bayonets were made from old materials, two thousand old India-pattern muskets were repaired and refurbished, five hundred old-pattern cavalry swords were cut, ground and polished. Three hundred and fifty iron wire masks, padded and bound for infantry and cavalry exercises, were produced. Eight suits of chain-mail and plate-armour were repaired and a thousand ancient swords, axes and maces were polished and refurbished. Leg-irons (178 pairs) were produced for service in Van Diemen's Land (Tasmania) and 177 iron trestles for barrack tables. It was not until 1829 that more orthodox work could be resumed: 477 infantry rifle barrels were bored and filed, 404 infantry rifle locks constructed and 2,353 gunmetal sights for naval ordnance were made. Lovell claimed an overall profit of £856 for the year.[4]

The work of George Lovell went far beyond that of a mere storekeeper or factory manager. In 1824 he began carrying out experiments at Enfield to prove the greater penetration of balls fired by percussion as against flint and steel. He had a new musket made on the percussion principle at Enfield in November 1831. Production costs, he thought, would be 42s to 45s, but if 10,000 were ordered the cost might come down to 35s. In January 1832, thirty muskets of the India-pattern had been converted to percussion at Enfield and forwarded to Pall Mall together with copper caps charged with 'fulminate of mercury'. Lovell was even beginning to convince the Army High Command of the need for change.

Lord Hill, Commander-in-Chief at the time, had been persuaded to admit that percussion had advantages in 'ensuring constant and certain fire'. Lovell pursued his investigations, having a recoiling target made for him at Woolwich and sent down to Enfield on a Powder Mills barge. By 1832 he had been in charge at Enfield Lock for 16 years. On the night of 1 April a riotous mob set fire to haystacks near the Gunpowder Mills at Waltham Abbey. The fire-engines from both government factories were called out, guarded by the West Essex Yeomanry Cavalry. The fires were extinguished after four or five hours but, in spite of the Yeomanry, the rioters managed to rekindle them many times. The yeomanry was required to remain on guard thoughout that year in an endeavour to arrest or deter rick-burning agricultural labourers, protesting thus blindly against starvation wages and repressive conditions. That year Lovell's windows were broken, along with those of many of his neighbours, in the celebrations at Waltham Abbey which followed the passing of the Reform Bill.[5]

The problem remained of finding enough work to keep the factory going. Indeed the closure of the Royal Manufactory was being considered in January 1832. This is indicated by the formulation of a list of officers, foremen and others who would become entitled to pensions in the case of such a contingency. A committee consisting of the Surveyor General of Ordnance and high-ranking army officers visited Enfield in 1834. It recommended that the employees should be retained on a six-day week, but that they should be paid for only five days, for it was asserted that 'the men, having no means of procuring other employment on Saturdays, would be left in idleness to brood over what they considered a grievance'.

A system had been worked out for requisitioning materials and fixing the prices paid for piece-work. All material required but not in stock had either to be requisitioned from the Tower or from other stations, or purchased from a contractor. On receipt it was inspected by the superintendent before being issued to the foremen, who then allocated the materials to the men as the work progressed. It was the duty of the foremen to make a daily report of the work done by the piece-workers under their control, no item being recorded until it had passed the viewer and had received his stamp. The workmen had to pay for damage done to materials by careless handling, the money being deducted from their earnings.

The men on daywork were mustered at regular hours, a roll-call was taken and a record of attendance kept. The superintendent was constantly passing through the workshops examining progress and watching that the dayworkers did not loiter. Towards the end of the week he collected the daily reports and records of attendance from the foremen and compiled a list of articles made and passed by each piece-worker up to Thursday night, allocating to each item its proper price from the price-book. From this list wages were paid by the storekeeper in person at eleven o'clock on Saturday morning, each man coming up to the pay-table and being called upon to state the amount which he considered due to him. This process must have demanded a widespread degree of numeracy. Every individual received his own money in cash and signed the list; the men were never paid in gangs. Daywork men were paid at the same time and in a similar manner. The arms were handed to the storekeeper following examination and the unused materials were returned to him. The arms were then sent to the Tower by the Waltham Abbey barges or by carrier's cart.

The state of Welch's Lane (Ordnance Road) thus became a matter of importance. The manufactory was not the only user, for the road was used by the servants of the Lee Navigation, also by farmers in the neighbourhood, thus its repair had never been considered to be the responsibility of the Ordnance, but Enfield parish was always unwilling to spend if someone else could be induced to pay. The Ordnance spent £206 to widen the road in 1837 and thereafter £25 a year was set aside for maintenance. Work on the construction of the Northern and Eastern Railroad had begun in May 1838 and Lovell was perturbed to find that the railway company had no intention of building a bridge to enable traffic to cross the line, it offered no more than gates with a watchman in attendance.[6]

The work-force in 1834 comprised the storekeeper, superintendent and clerk and forty-three employees, including four foremen and viewers, thirty-four artificers, first to third class, a day-warden and four labourers. Top wages were 39s a week, labourers being paid 2s a day and upward. The nucleus of the work-force had been reduced in some branches to the lowest possible point, and this rendered the working of the factory uneconomic. There was a single band welder for instance who, if employed in full profitable work in his own trade, would forge 70 musket barrels a week, but this would then require three band filers, each with an assistant, 17 lock filers to make the locks for them, 5 stockers to put them together, 7 screwers-together to finish them, besides polishers and other workers. A report of that year nevertheless stressed the importance of retaining the nucleus of the manufactory in all its parts in time of peace. Thus the men had to do whatever work turned up, the band welder, for example, made other articles in iron required by the department at large.

Lovell urged that high pay should be maintained, stating his opinion in confidence that the current rate of pay was lower than it ought to be 'by the amount of the late reduction'. That reduction had been recommended by Lovell himself but only as a lesser evil, as against stopping the men entirely on Saturdays. Saturday closing, he held, situated as the manufactory was with regard to locality, would have been very injurious to its interests. He suggested that members of the select committee (ie to enquire into working establishments at the Enfield and Waltham Abbey factories), should 'see the workmen and state to them personally the imperious necessity under which all the departments find themselves to reduce public expenditure. Such a communication', he went on, 'would do much to allay feelings of discontent and to remove from the officers' (including of course himself) 'part of the odium ... from being the organs of unpalatable measures'. Obviously serious discontent existed at Enfield Lock. Skilled artificers felt it was a grievance that, when they were employed on work outside their own trades, they were paid by the day and so could not exploit their skills as they would when engaged on piece-work.[7]

Lovell began experimenting with percussion soon after his appointment. In November 1831 he had an India pattern flintlock converted at Enfield to the percussion principle. It was sent for examination to Lord Hill, the Commander-in-Chief, who admitted that it had some advantages in 'ensuring constant and certain fire'. Lovell was concerned that the tests for percussion were inadequate and sought permission to make changes. To facilitate these he had a boarded enclosure constructed to carry out his experiments. In December the following year he had six carbines for the cavalry altered to the percussion principle, they were forwarded to Pall Mall by carrier. Enfield continued to be closely involved in the changes going forward and 300 flintlock muskets were sent by barge from the Tower for conversion, in batches of 40 a week, from December 1834. By February 1836 Lovell had achieved a pattern for a rifled percussion musket with which he could be satisfied. It was a completely new weapon, constructed on the cap and nipple plan for simplicity, cheapness and security. The barrel was rifled with eleven grooves instead of seven. He had provided a raising leaf-sight for ranges over 350 yards, with the 'V' narrower for greater accuracy. He estimated that the cost would be £3 10s as against £3 1s 1d for the flintlock. Major Dundas was enthusiastic. 'Mr Lovell's new rifle' he said, 'seems to possess all the essential requisites to render it efficient for military purposes'. Lovell was ordered to send the pattern, for tests on active service, to the Rifle Brigade stationed in Canada. The choice of weapon 'for military service', wrote Lovell, 'must turn upon the convenience and facility of use in the hands of soldiers ... trials should be made by men of the Rifle Brigade,' the reports brought back were favourable. Converted carbines were sent to the cavalry whose officers proposed minor improvements which were then carried out at the factory. Meanwhile Lovell continued to search for improvements. He had secured a rifle from the Duke of Brunswick the barrel of which had double grooving and the ball fitted exactly. This, said Lovell, gave it advantages, for the barrel would last much longer and the rifle would clean itself at each time of firing. He was now preparing a model for the Rifle Brigade and also experimenting with a longer carbine. He sought permission to set up six rifles of each kind, with long and short barrels, and double grooved, to compare the Brunswick with his New Pattern Percussion musket. These were sent to Woolwich in November 1836 and a committee of artillery officers was set up to carry out tests. Lovell advocated that the bore of all military small-arms should be standardized at .753 of an inch, as in his percussion musket, so that one ball should fit all the arms in service. The number of different calibres had been much reduced since the war and he urged, now that the whole of our small-arms were about to be changed to the percussion principle, standardization ought to be achieved at the same time.[8]

An opportunity to expand production came in 1837, for Lovell's percussion musket was finally approved by the Commander-in-Chief in July that year. By this time moreover the store of arms in reserve had been reduced to such a point as to make it imperative that it should be replenished. 'It ought now to be decided', wrote Lovell, 'how a regular supply of improved arms is to be obtained in the future'. He stated that it was his conviction 'that the Government of this country can never depend wholly upon private trade', at the same time he thought that it would be imprudent for all the work to be done in government factories. So few orders had come into the Royal Manufactory in the years up to 1837 that it had not been possible to train young hands. At the same time government orders to private firms had been so intermittent that manufacturers had not been able to retain highly paid artisans. Thus a shortage of skilled workmen had been created, there being no other customer for military small-arms except the Army and the East India Company. It now became necessary to secure further skilled labour if the factory was to participate in the re-equipment of the forces with percussion weapons, and the availability of housing on the site must play an essential part in this. Lovell therefore felt that it was necessary that the cottages should be properly maintained, and in February 1836 £198 was spent to render habitable cottages lying empty and 'now suddenly required for additional workmen'. In 1838, glass window panes in three of the cottages, broken by the tenants who had had to move out, were replaced and the interiors were whitewashed before new families moved in. Four of the one-storied cottages were given an upper floor in 1839.

Most arms had for a long time been manufactured in Birmingham, even swords, although the steel could be produced only in Sheffield. Barrels and locks had been made entirely in Birmingham where all the parts were prepared by individuals in 'the garrets and lodgings of the working population'. The locks, stocks and barrels were each produced by individual tradesmen by whom the making of each part would be allocated to a separate worker, for example each various component of a lock would be made by a different man. When machinery was required, as in boring or polishing, steam power was hired by the week, even by the day. Women and children could be seen at all times hurrying through streets bearing parts between workshops, or to a warehouse, much as had been done in the time of Charles II. The merchants would then deliver them out again in sets to a

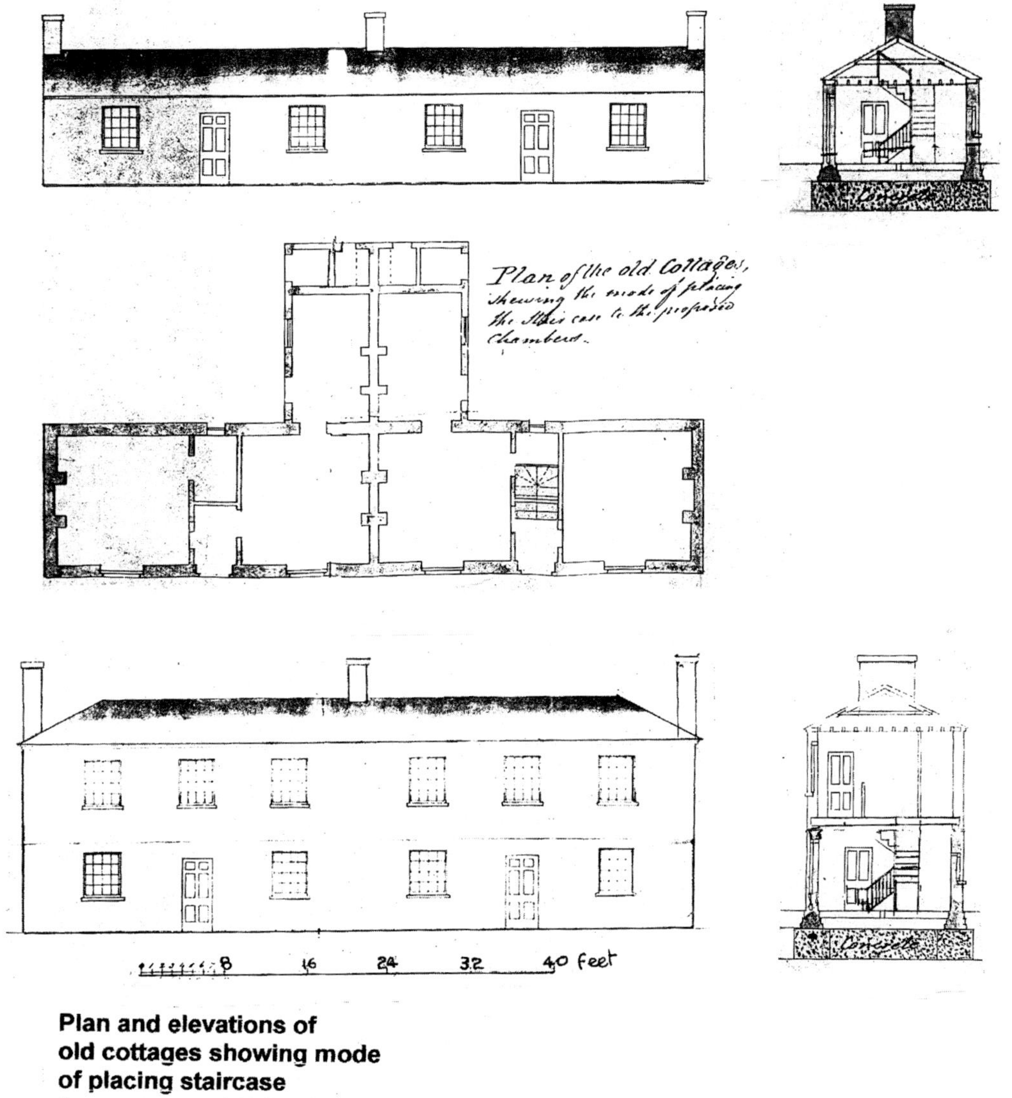

12 Some single-storied cottages for workmen were given an upper floor in 1839

different class of workmen who would set up the parts into guns in their own rooms or shops. A very limited degree of accuracy was required, and uniformity of the parts was not demanded.[9] By thus keeping separate the contracts for parts and the contracts for setting up the arms, Ordnance gave the trade no incentive to install machinery for the production of standardized parts.

Lovell strongly advocated that, now percussion had been adopted, all services should be provided with the new weapons. This would require 140,000 arms, moreover he considered that five years' production should be stored, making the total required 198,000. He strongly urged that Enfield Lock should play a major role in the supply and the situation brought about a modest expansion. The building containing the barrel and lock shops was extended upward to provide space for the rough stocking and finishing branches and a new building was constructed to provide offices, a warders' room and a room in which the men could assemble on pay-day. Above this was provided a long room for a pattern collection. This was in 1839, before the disastrous fire at the Tower in 1841is said to have caused the removal of the pattern collection to Enfield. A new coal yard was constructed capable of storing 300 tons, with space for weighing, and the former coal store was converted into a store for seasoning the gunstocks which the Board had purchased, the work cost £4,043.

Faced with the rearming of the entire forces with percussion arms the Board had to consider whether to achieve this mainly by private contracts. The alternative would have been to establish a manufactory on a scale sufficient to carry out the work themselves by machinery. Henry Wilkinson was one who advocated this, setting out the advantages in costs, uniformity, and diminished dependence on skilled labour. He maintained that machines existed in this country which could be adapted to make military small-arms. But the Board at this time was faced with no crisis which would propel them to take such drastic measures.

Lovell sought permission to order four tons of iron for the making of 1,000 barrels by rolling. He recommended purchase from George Adams at Wednesbury for he had found that the iron produced in Birmingham was impregnated with sulphur and unfit to be used on the new pattern percussion muskets. He tried to interest the Board in the purchase of two compressing machines at a cost of £1,000 from Napier of York Road, Waterloo Bridge. These were capable of producing musket balls in quantities sufficient even for wartime, at a rate of 80 balls a minute. The machines were purchased in April 1839 and were installed in the Royal Laboratory at Woolwich.

Enfield still seemed unable to undertake manufacture on any large scale. Work was held up on 2,000 infantry rifles ordered in 1836, the pattern had yet to be agreed and this delayed production. Further delay was caused by the bursting of a large grindstone in October 1837. The grindstones were powered by the watermill where a horizontal shaft rotated two hollow cylinders. Bands passing over these cylinders rotated the Derbyshire grindstones, grinding barrels in a shed attached to the ground floor of the mill. The large stone, which weighed 4½ tons, was impelled at a speed equal to 3,600 feet a minute; on bursting it destroyed the shed and everything around it. Fortunately no lives were lost but much of the work being done at Enfield had to be re-allocated to the trade. Grinding was a pernicious trade. In the winter of the year 1839 hundreds of rammers had to be altered for use on the new percussion muskets, 5-800 a week. The work had to be halted because of the serious illness of two of the grinders

13 It was planned in 1840 to convert the view room and store (2) entirely for storage and to build a new view room (10). A crane to lift 5 or 6 tons was to be installed by the basin (11). The new offices shown (7) probably housed the pattern room collection

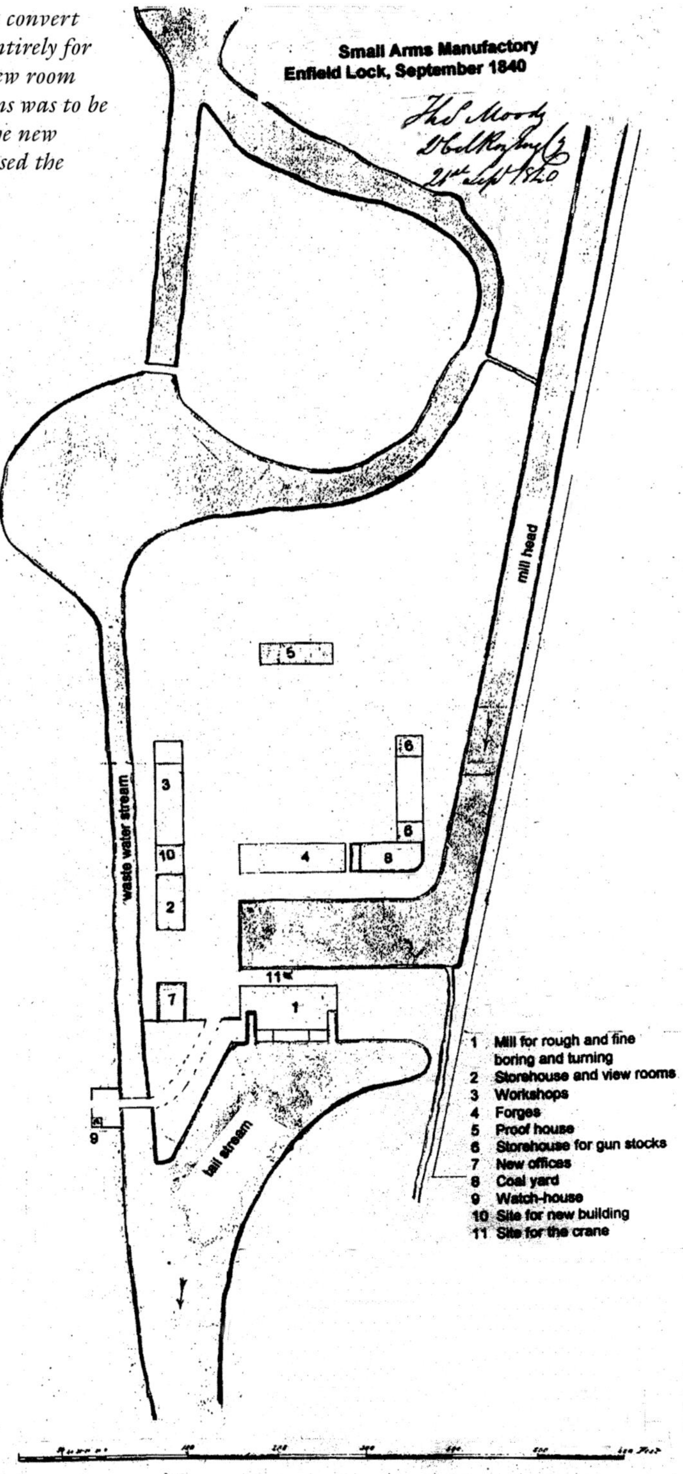

working on the grindstones in the cold wet weather. One it was feared would never work again.

Lovell remained of the opinion that the barrels, rammers and bayonets should be purchased only in Birmingham, locks, he thought could be produced both at Enfield and at Birmingham. The setting up should be divided between Enfield, the London gunmakers and the Birmingham manufacturers. The Enfield factory remained essential in ensuring that a body of workmen capable of doing military work was retained in this country. Birmingham remained, Lovell admitted, 'the great school in which the Royal Manufactory must always seek its hands'. At this time Lovell and Gunner went to Birmingham to open Ordnance premises there.[10]

It was ordered, 3 June 1839, that all parts henceforth be purchased from the trade in Birmingham. Over the ensuing year or two the difficulty in obtaining supplies of percussion arms gradually disappeared, 4,000 were in store at the Tower by January 1840 and it was confidently expected that there would be 10,000 by the end of March. There were now 14 contractors for setting up arms in Birmingham and 14 in London. The Board hoped that they might compete one against the other, the Ordnance factory at Enfield being used as a check on their charges, but the spread of railways facilitated liaison between the two groups of manufacturers. They colluded repeatedly, and when criticized they failed to fulfil their contracts and blamed Lovell, claiming that his inspection was too stringent. Lovell continued to urge that the Royal Manufactory be permanently enlarged to provide room in the workshops to extend operations in case of an emergency.[11]

George Lovell was appointed Inspector of Small Arms in 1840, following the retirement of Manton the Master Furbisher and, following the fire at the Tower in 1841, the Pattern Room there was transferred to Enfield. Lovell had always stressed the importance of strict inspection of the weapons produced at Enfield, but now his function, as Inspector of Small Arms, put him in a position to impose the same standards upon all small-arms produced for the Government. He insisted that inspection should no longer be done in the workshops of private tradesmen and the trade reacted with hostility accusing him, in a pamphlet of 1843, of imposing a 'vexatious view'. Lovell responded, before the 1849 committee (House of Commons Select Committee on Army and Ordnance Expenditure), to bitter complaints concerning the minute sub-division of inspection. He declared himself glad to hear of the manufacturers' dissatisfaction because it indicated that inspection was being carried out according to his orders. He urged that viewers should be well paid so that they would not be enticed to take bribes, they should receive an annual salary and should be available to work either at Enfield, the Tower or at Birmingham. Eight viewers were employed at the Tower in 1841, eight at Birmingham and six at Enfield. The Enfield viewers were men with long experience; John Smith, the principal viewer, had been employed by the Board of Ordnance since 1804 and had held his present office since July 1815, John Jackson had been appointed barrel-viewer in 1809, Francis Rowley, sword-viewer, in 1823,

Joseph Honnor, roughstock-viewer in 1834, Abraham Greene, finishing-viewer, in 1836 and Charles Dure, lock-viewer, in 1839. Their pay in 1840 was 6s 6d a day, it had not kept pace with the advance in wages in the gun trade generally. A petition for an increase, strongly supported by George Lovell, was granted and from February 1841, their pay was to range, according to length of service, from £110 a year to £130.[12]

Also in 1840 Gunner was appointed storekeeper. Lovell thus became responsible both for the overall direction of the factory at Enfield and the supervision of the Department of Ordnance small-arms workshops in London and Birmingham; secondly he took control of all orders for small-arms issued to contractors. He now sought to expand the factory at Enfield and over the ensuing years production increased from 7,000 small-arms and 1,500 swords a year in 1840, to 50,000 muskets and 5,000 swords by 1853. Many were produced for sale abroad. It had been proposed in January 1840 that a new barrel welding-mill, where the barrels would be drawn through rollers rather than hammered and welded, should be built at a cost of £26,412. This was intended to bring barrel making in Enfield up to the standard of the most advanced practice in Birmingham where the introduction of the process had caused riots among barrel-forgers. Construction work was begun immediately, but by May the work had been halted. Barrel-rolling capability was not installed at Enfield until 1853, at a time when war with Russia looked inevitable.That August (1840) Gunner asked for an adequate storeroom, he complained that articles to be issued out to the work-force were dispersed in various workshops and in the proof-house. He also stressed the need for a crane to lift five or six tons for unloading heavy stores at the mill wharf, particularly the large grindstones some of which weighed three tons and had to be unloaded by men with lifting jacks and levers at the risk both of injury to the men and damage to the barges. A plan was eventually agreed to convert the existing view-room and store entirely for storage and to erect a new view-room. The work would cost £1,612.

The pattern of a new line musket was approved on 27 October 1841. Three days later the workshops at the Tower were destroyed by fire together with 280,000 guns including most of the percussion weapons of the 1838 and 1839 patterns. New contracts had to be issued and the weapons now produced were classified as the 1842 pattern.[13]

The year 1842 had been the summit of Lovell's career. Having reorganized the ordnance factories, he had recently taken upon himself, without any instructions from his Board, to act as arbiter between the contractors and their men in the wage negotiations which took place that year in London. The Birmingham men had sent representatives but, as he proudly told the 1849 Parliamentary committee, 'they left it to me and there have been no strikes since'. Firearms recognized as Lovell's design were now widely used. His son Francis was appointed Assistant Inspector of Small-Arms on 10 March 1843. Lovell however had made many

14 For much of the year the Sewardstone silk mill was virtually inaccessible either by land or water. Here, from 1806, many pauper children from Enfield were incarcerated

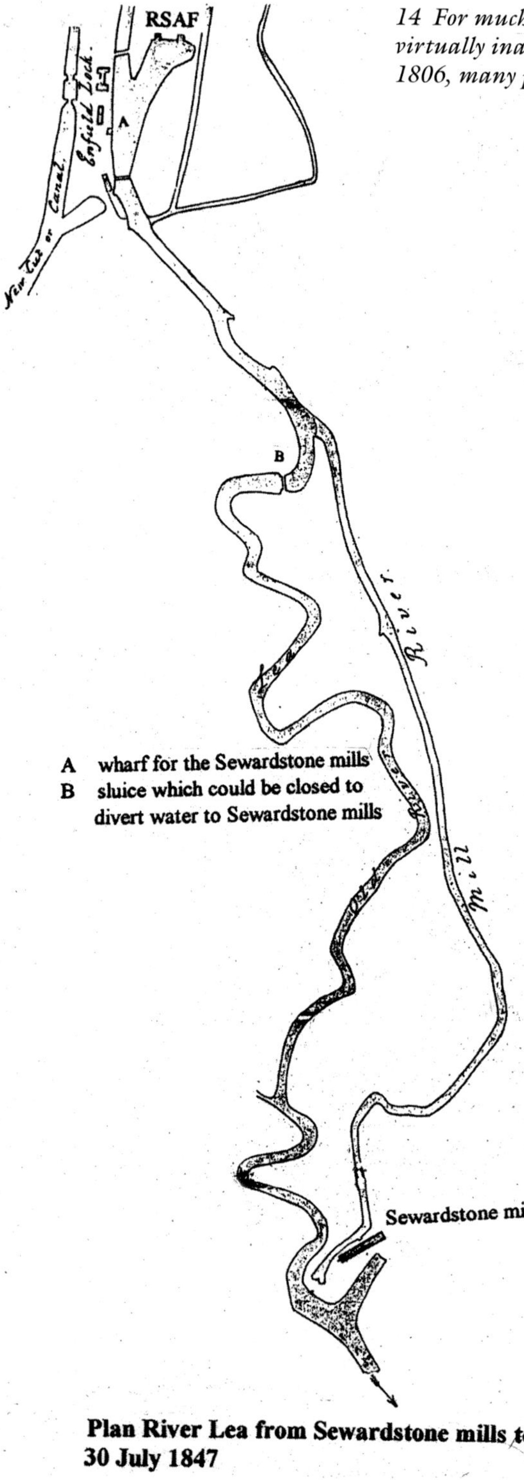

Plan River Lea from Sewardstone mills to Enfield lock, 30 July 1847

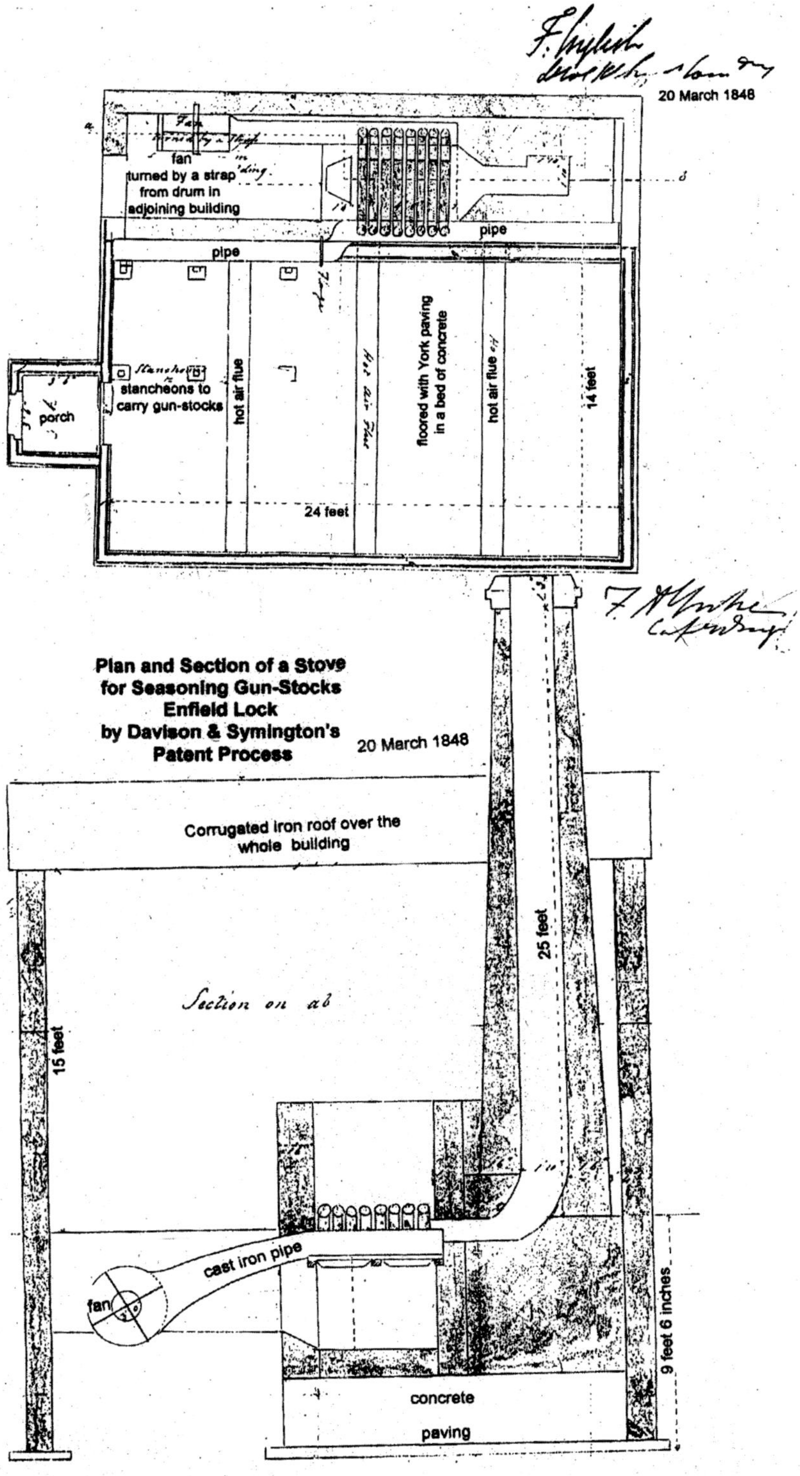
20 March 1848
fan
turned by a strap
from drum in
adjoining building
pipe
pipe
stancheons to
carry gun-stocks
porch
hot air flue
floored with York paving
in a bed of concrete
hot air flue
14 feet
24 feet
Plan and Section of a Stove
for Seasoning Gun-Stocks
Enfield Lock
by Davison & Symington's
Patent Process
20 March 1848
Corrugated iron roof over the
whole building
Section on ab
25 feet
15 feet
cast iron pipe
fan
9 feet 6 inches
concrete
paving

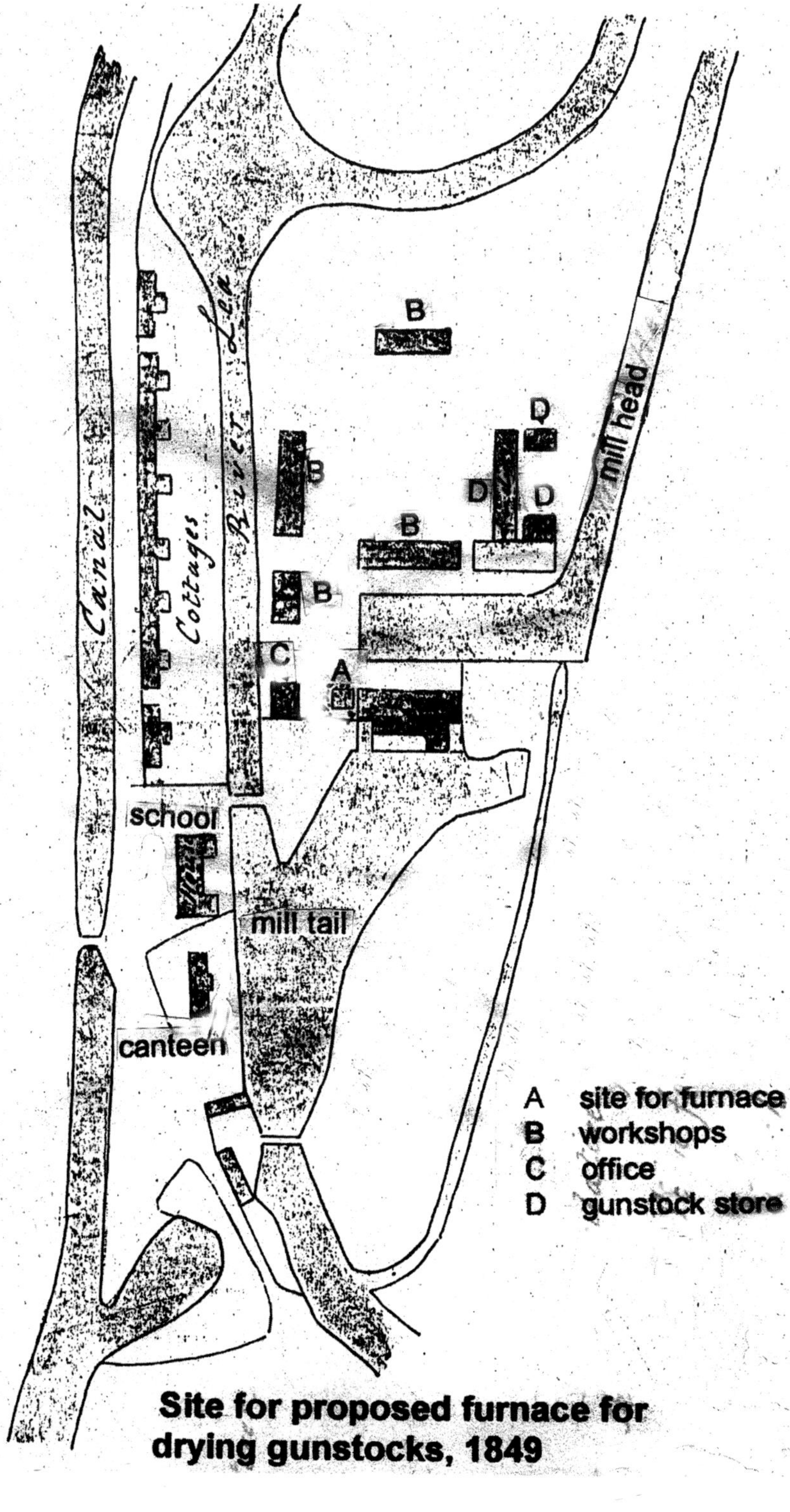

Site for proposed furnace for drying gunstocks, 1849

enemies. Unsuccessful small-arm designers blamed him for the rejection of their designs and he was bitterly attacked, especially by William Greener the gunmaker and Colonel Peter Hawker, both resented their alleged humiliation by him before the Select Committee of Officers. Lovell faced his critics with considerable resolution, even daring to contradict the Duke of Wellington, the Commander in Chief, over which method of attaching the bayonet to the musket should be adopted. Wellington had been asked to choose one of four methods and, having made his selection and revealed it to his underlings, he was outraged to find it rejected by the Board in 1844 at the instigation of Lovell and in favour of a fitting designed by Lovell himself. The Duke's resentment was transmitted indirectly to the Board of Ordnance, 'the Commander in Chief had hoped that this important subject was finally disposed of by the selection of bayonet number one, which His Grace had made .... The Commander in Chief understands now that Mr Lovell considers it liable to many objections ... among others that of expense ....' Wellington grudgingly accepted the decision but it is unlikely that he forgave the offence to his dignity.[14]

Lovell had begun to introduce more advanced machinery at Enfield Lock; a lock-plate driller in 1842 and percussion machines in 1845. A strike of gun-stockers throughout the country, which began in 1845, became more bitter as it became more protracted. Because of the strike the store of gunstocks had fallen very low by 1848, both in London and in Birmingham and stocks had to be purchased on the Continent. Lovell now sought permission to order 15,000 from the Patent Dessicating Company at £12 10s a thousand. Meanwhile a dessicating room was being built at Enfield, 24 feet by 14 and 12 feet high. The number employed at Enfield had risen from the time of the adoption of the percussion principle in 1837 and the work-force continued to grow until the industrial unrest of the mid-forties, when growth was halted. In 1848 the safety of the two government factories on the River Lea became a matter of considerable anxiety. This was at the time of the presentation of the third great Chartist petition on 12 June, the West Essex Yeomany had to be called out. Growth was resumed in 1849, wages paid out that year amounted to £2,200 compared with £1,400 in the previous year.

The gun trade continued to obstruct at every opportunity Lovell's drive for efficiency. By 1848 there were eight contractors for barrels and ten for locks, they had become well organized and had secured powerful friends. They showed little enthusiasm for completing government orders yet protested violently when the orders were cancelled. In February the following year the Board set up a system of competitive tender, but the situation did not improve. The conversion of flintlocks to percussion, for service in the colonies, was so badly done in Birmingham that Lovell had to install machinery to do the work at Enfield. Lovell's musket came increasingly to be criticized and he toured the Continent in search of a possible replacement. He found a promising design in the Minié rifle manufactured in France and Belgium. In May 1851 he was instructed to construct dies

PLAN OF THE R.S.A.F. AT ENFIELD LOCK 1851

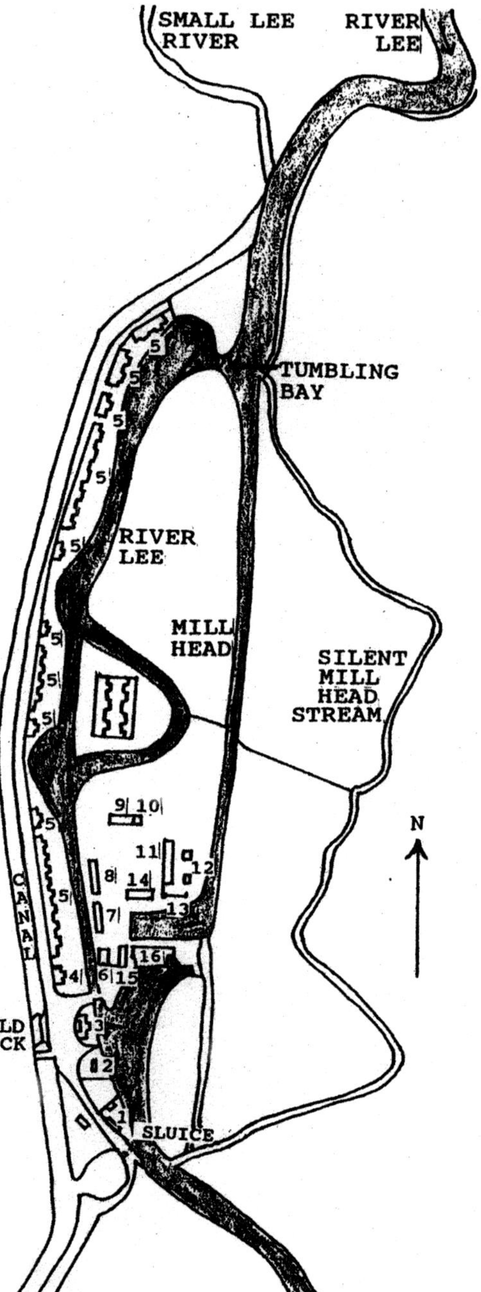

KEY

1	STOREKEEPER AND SUPERINTENDENT (1 PERSON)
2	CANTEEN LET TO JOHN MARTIN
3	SCHOOLS AND CONGREGATIONAL PURPOSES
4	CLERK TO STOREKEEPER
5	59 COTTAGES
6	OFFICE AND PATTERN ROOM
7	STOREHOUSE AND VIEW ROOM
8	WORKSHOPS AND TWO STORES
9	PROOF HOUSE
10	FIRE ENGINE HOUSE
11	STOCK STOREHOUSE
12	IRON SHEETS
13	COALYARD UNCOVERED WITH BRICK WALLS
14	FORGE SHOPS
15	DESICCATING CHAMBER FOR MUSQUET STOCKS
16	MILL

SCALE 0 200 400 FEET

and tools for making Minié bullets of an elongated shape (instead of round) which would give greater range and an increased rate of fire. He was also asked to build four copies of this Belgian pattern rifle at Enfield, and on two of them he changed the rifling. Once again he faced opposition from the Duke who refused to sanction any reduction in the calibre of the bullet, thus the 1851 pattern rifle had to be unnecessarily heavy. Nevertheless it was adopted as the standard infantry arm and orders for 23,000 were issued to Birmingham contractors, but a dispute immediately arose between the operative gunmakers and their employers. Lovell was forced to go to Birmingham to seek a settlement, but he was to receive little thanks for his intervention.[15]

The series of misfortunes which marred Lovell's later years had begun with the death of his son Robert in 1845 while purchasing gunstocks on the Continent. The following year there had been a strike at Enfield and one of the workmen had made a murderous attack on him. The year 1847 had seen troubles with the contractors who had never forgiven him for the strict standards which his viewers had imposed on their work. Even the workmen, whom he had attempted to protect from their employers, became hostile to him. His downfall was brought about by the indiscretions of his son Francis, the Assistant Inspector of Small-Arms, who was accused of falsifying the accounts for gunstocks.[16]

It was in December 1851 that the Store Accounts Office called for an explanation as to a large deficiency of musket stocks. George Lovell was visited at home where he was lying ill and bereaved, having recently lost his wife. The deficiencies were demonstrated to him but he could only refer the matter to his son Francis, suggesting that the explanation might lie in a clerical error. Francis Lovell however was unable to explain and the conclusion was reached that 6,452 stocks had been paid for in excess of those delivered, and that Messrs Pirlot and Simons had been paid £1,628 more than their due. Although Francis had not gained a penny on the transactions he was ignominiously dismissed. George Lovell did not escape the humiliation which had descended on his son. He was criticized for not having himself brought the case before the Board, and his conduct was condemned as 'wrong and reprehensible'. Nevertheless in consideration of his 'long and useful service in the department' it was allowed that he should be only cautioned.

'Long and useful' was an ungrateful and ungracious way to describe Lovell's invaluable work over so many years. He wrote a humble letter to the Board pleading for mercy on behalf of his son, on whom, with his wife and young family, dismissal would bring utter ruin. His appeal was disregarded, the dismissal was not rescinded and the Board proceeded to impose further difficulties on the stricken family. Lovell was now ordered to take up residence in Birmingham and to manage the London side of the business by visits to the Tower and Enfield; he had to dispose of his house and find another in Birmingham. George Lovell retired in May 1854 and at the time of the 1854 Parliamentary Commission he lay seriously ill at home. He died soon afterwards broken in health and spirit.[17]

18 Barrel-rolling machinery was delivered in March 1853

Lovell had long been in favour of improved machinery at Enfield. A Nasmyth steam-hammer was installed in 1851 and a rifling machine of Lovell's invention, and in 1852 a planing-machine was purchased. In a letter written in March 1852 he had complained that the Ordnance had been paying £42 a ton for iron rough rolled into musket barrels at a time when iron cost no more than £18 a ton. This he said was because the department had not had the means and machinery for making the barrels at Enfield, although abundant water-power was available. The barrel-rolling machinery was delivered in March 1853, but no labour could be spared for its installation, for the men at this time were fully engaged producing weapons. The Royal Engineer's Department was therefore called in to erect and fix the machinery and was required to do so without halting the mill. From that time the factory was able to make 200-250 barrels a week. The process of rolling barrels required the oversight of a man highly skilled in the process, and such a key worker had been induced to leave Millward's at Birmingham, without notice, to come to Enfield. Barrel rolling was at first done on day-work but only temporarily.

The barrel borers were on piece-work and their pay had to be above that normally paid because there was still insufficient work for them to earn their money. James Gunner did not think that any great savings would result from the application of machinery to stamping locks, which had always been forged at Enfield by hand labour and tools. The cost of forging each lock was 6d, stamping might be done for 4d, but he thought that forged work was to be preferred. A fair margin was to be saved on stocking by machinery, he said. Under the current system of assembly, if he wanted a hundred guns in a week, the process was so labour intensive that he would have to employ a hundred men with the boys under them, for only one gun per man could be completed in a week. The stock was taken in a rough state, the barrel let in, then the lock, then the gun was rounded up and the bands put in. Following this another workman would have to screw the assembly together and let the mounting in. Walnut wood for stocks had become scarce in this country. Some stocks were delivered to a store at Weedon (Northants) where they could remain to dry for a year or two, although stocks cut nearly green from a plank could take as long as three years to season. When stocks were required in an emergency the factory now had the equipment for dessicating by hot air; by this process stocks could be made ready for use in four or five weeks.

The Board of Ordnance in its quest for a new rifle was faced with a choice from a number of patterns, none of them completely satisfactory. It was therefore decided to recommend a theoretical rifle embodying the best features from each. A new bullet of an elliptical shape was also recommended. Two of these rifles were made at Enfield and work on the 1851 pattern was suspended. A series of trials was held at the Royal Manufactory in 1852 on these new weapons and they were successfully tested in December 1853 and thus the famous 1853 pattern rifle was adopted.[18]

The Crimean War had become inevitable by the end of the year 1853 and it was launched on a wave of patriotic anti-Russian fervour. The Allied armies landed in the Crimea on 14 September 1854 and on the 17 October the siege of Sebastopol began. The war created an urgent demand for muskets. Once again, as at the beginning of the Napoleonic War, the country discovered that there was a crisis of supply. Henry Wilkinson had long ago advocated the employment of machinery for the manufacture of military small-arms. In May 1837 he had set out the advantages in cost, uniformity, and the ability to employ unskilled labour. He had pointed out that various machines, used in this country for other purposes, could readily be adapted to the manufacture of small-arms, he had been willing to oversee the provision himself, but his urgings had passed unheeded.

The system for the supply of small-arms from private contractors was severely criticized before the Parliamentary Commission of 1854 (set up to examine the most efficient mode of providing small-arms for Her Majesty's service). All the component parts for the production of muskets were procured by open contract and after inspection were placed in store. A small proportion of the parts were then issued to Enfield to be set up and finished, but most of the materials were delivered out to contractors who returned the muskets to the Board completed. Lord Raglan, the Master General, complained that this process had been operating badly, he held that supply ought to be made independent of contractors. Admiral Sir Thomas Hastings, Surveyor of the Ordnance, agreed. It was recalled that in May 1851 the Board had entered into contracts for the supply of 28,000 rifled muskets by February 1852. A few days before the tenders were returned, the workmen submitted a demand for increased prices for piece-work, the muskets were not delivered until November 1853. Contracts for 2,000 artillery carbines were entered into in April 1853 and the order still had not been completed in September 1854. In August 1852 contracts had been sought for 20,000 rifled barrels. Barrel rolling was one part of the trade where machinery had been successfully applied, but the tenders received from seven contractors had been identical at 14s and, although they were advertised again throughout the United Kingdom, only the same parties responded, and all of them offered a price of 13s. Then came the change to the 1853 pattern. Eventually the work was delayed so long that the setting up could not be completed until September 1854, the month our troops landed in the Crimea. Even worse, contracts were agreed on 6 October 1853 with eight manufacturers for the production of 20,000 locks but by 17 January only 755 had been delivered. It was complained that the workmen took the opportunity to strike immediately their firms received government orders. The Commission found it difficult to understand why the gun trade in this country, being so large and being an exporter to other countries, should find it so difficult to meet the orders of their own Government. Mr Clive and other barrel rollers in Birmingham had the capacity to roll 300,000 barrels a year, but they constantly failed to fulfil government orders. Undoubtedly it was because they could find more remunerative employment for their machinery. The delays of the manufacturers were not without consequence for the men serving in the Crimea. Captain Johnny le Couteur, commanding a company of Coldstream Guards, complained of the incompetence of his headquarters, the lack of transport, 'no mules or even carts', he then went on; 'the small-arms also are in a shameful condition... instead of armourers to look after them as they came in, they have been allowed to rust'. They were in any case of an inferior pattern. 'The Enfield is a far better arm', he wrote.[19]

Hastings in his evidence before the 1854 Commission read many letters from contractors endeavouring to excuse their delays; they cited combinations and strikes, difficulties in getting coal, frosts, illness of skilled artisans and accidents to machinery, but the inherent incompetence remained obvious. The Army urgently required 90,000 Enfield 1853 pattern rifles and, when all units were eventually equipped, 30,000 replacements annually would be needed. The Board of Ordnance now came to the conclusion that it was essential to take the whole manufacture into its own hands. An establishment had to be planned to turn out 500 muskets a day.

Lord Raglan and Admiral Hastings were impressed by the report of the Commissioners who had visited factories in the United States where muskets were made by machinery. They had also received advice from John Anderson, the chief engineer at Woolwich Arsenal. The advantages of machinery were said to lie in the reduction of costs, (costs in this country, for a musket, comprised 17 per cent materials, 83 per cent labour), in the production of uniform interchangeable parts, and above all in the speed of manufacture. It was reported that the United States had two factories each capable of producing 30,000 muskets a year. Anderson recommended that machinery should be purchased in the United States.

Sir Joseph Whitworth, an eminent engineer and a manufacturer of machinery, gave evidence that the US government establishment at Springfield had an ingenious system for the manufacture of gunstocks which could be completed in the space of 23 minutes, nevertheless he expressed doubt whether the parts would fit together. Walls, the headmaster of the Government School of Art at Birmingham, on the other hand, reported that he had seen the Americans put locks, barrels and furniture into a gunstock, though each part had been taken indiscriminately. Nasmyth, the inventor of the steam hammer, held that the introduction of machinery into the gun trade was certainly feasible. Another witness, Captain T. Hastings, believed that a musket for which the contractors were paid £3 7s 3d could be made by machinery for £2. The Government had been offered machinery many years ago, he said, had they purchased it then and not let it go elsewhere, £1,500,000 could have been saved. Although Colonel Colt had introduced machinery into this country to produce pistols, Hastings knew of no example in England of the application of machinery to the making of muskets, except at Braziers for lock making, and at the government factories, which had set an example, both at Enfield and at Birmingham. He had warned the trade that, unless it improved supply and reduced prices, the Government would be compelled to purchase in Liège or even in America. This latter suggestion had aroused hilarity among the contractors but he pointed out to them that supplies could be brought over by Cunard's vessels as easily as from Yorkshire. Further he had indicated to them the possibility that Government might itself embark on full-scale small-arms manufacture. Anderson had drawn up a plan with all the machinery aligned in neat rows in one large room 180 feet by 500. The barrels, bayonets, locks, stocks and other parts would be moved across the room at each stage of production, reaching the far side in an almost finished state. From there parts would be conveyed by rail, the barrels for example to the proving room and thence into a store for examination. On the side of the large room would be a brass foundry, a dessicating furnace and a boiler house, arranged so that one chimney might serve all three.[20]

Enfield was considered unfavourably as a site for this projected factory, indeed Anderson's proposal was that the new factory should be built at Woolwich. Housing was very difficult at Enfield Lock for as yet no land companies had set out estates in the vicinity. Sixty houses existed on the site but there would be no room to build more since the land would be required for workshops. There was no local community, the nearest village was at Waltham Abbey two miles away, where there was a market. Ponders End was two miles away, and Enfield Town four miles distant. Sir John le Couteur, from Jersey, who visited Enfield Lock at this time, later wrote; 'the Government must have been at its wits' end for a locality to come to such a marsh ... in which to manufacture small-arms. It is the most marshy spot in England that I know'. Skilled men did not relish the idea of leaving Birmingham for Enfield and it often happened that, even when they had been persuaded to come, they would ask to return home after five or six weeks. In the back of their minds was always the fear that should they be laid off they would be unable to find alternative work in Enfield. The Enfield factory lay several hundred yards from the railway station and everything had to be carted. One witness before a parliamentary committee many years later, in 1887, offered an alternative explanation of how the new factory came to be built at Enfield. 'It is altogether an uneconomical place and it was felt to be so by those who ordered the factory to be erected,' he asserted 'the trade was cunning enough to see', he went on, 'that if the factory was erected at Woolwich, it would be a serious nuisance to them, and it was owing to the efforts of the trade that it was erected at Enfield'. In support of his argument he quoted Mr Goodman the chairman of the Birmingham Small Arms Co. He also quoted a letter dated 26 May 1855 from Lord Panmure to Mr Scholfield, Member for Birmingham, explaining that, 'by instituting the factory at Enfield it is by no means intended to pass by the gun trade, the object being to enable the Government to keep up a moderate supply, and to secure the public from any combination of trade ....' Later secretaries of state, Lord Herber, Cardwell and others up to Childers, had agreed with the principles laid down by Panmure but, the 1887 witness went on, 'the facts had proved too strong for them'. For the needs of war created an enormous demand for the 1853 Enfield rifle, 'moderate supply' proved insufficient and the factory at Enfield Lock had to be built on a scale large enough to produce 100,000 a year.

A fundamental change had already taken place by 1854 at the Ordnance Factory and it had aroused strong hostility among the contractors. Enfield, from being a mere repair establishment, had become a considerable manufactory. Large quantities of locks and barrels were being made there. Witnesses complained, before the 1854 Commission, of the lower prices being paid to the Enfield workmen, because of which, they said, it was not improbable that all government orders would soon be supplied from Enfield. John Anderson thought that much more could be done in Enfield to increase production, a large number of machines were needed, both for musket and rifle making, indeed he averred that nothing less than the introduction of self-acting machinery to produce standard parts to make up the gun could give complete satisfaction. His reputation was already established for he had carried out a considerable reorganization at Woolwich and the re-equipment of the Royal Gunpowder Factory.

The Royal Armoury Mills was still powered entirely by water from two 45hp water-wheels, 9 feet wide and 18 and 20 feet in diameter. One was used for rough-boring and grinding, the other for fine-boring. There was a supply of water running to waste at some seasons of the year and the work was seldom stopped for want of water. The largest number of rifled muskets the factory could turn out with the present power was 15,000 to 20,000 a year. Gunner was of the opinion that the advantage was with steam which could more easily be regulated. When new weapons were introduced, three copies of the pattern were made at Enfield, one for use in the factory and duplicates for the Tower and Birmingham. Gunner proposed that each contractor should be allowed to use a pattern for a few days only, while he made his gauges.

Government arms factories abroad, like that at Liège, were mostly managed by artillery officers but the status of the works at Enfield had never been of sufficient importance to justify the appointment of an officer of senior rank. Although there was now a great deal more work done at Enfield than formerly, it was not possible to place army officers under the control of civilians. The men were employed from 6am to 8.30pm They were kept on indefinitely, providing their conduct was good, and after a great many years a few of them got a retirement allowance, but most of them died in harness. Gunner said that he could not remember that the men had ever gone on strike and he claimed that although their pay was a little less than their counterparts at Birmingham, their employment was continuous.[21]

Notes to Chapter 2

1 WO 55.753, WO 44.679, 680, Blackmore, *op cit* 272, 273, SUPP 5.689
2 WO 44.682, WO 55.754
3 WO 44.682
4 Blackmore, *op cit* 273, PP 1830 xviii, 308-9, WO 44.677.772/3
5 WO 44.132, 677, WO 55.754, G.O. Rickword, *The West Essex Yeomanry Cavalry*, 1830-77
6 SUPP 5.689, WO 55.754, WO 44.132
7 SUPP 5.689
8 WO 44.677, 742, 676, 654, 652, 649, 627, 684/5, 624
9 WO 44.133, WO 55.754
10 WO 44.133, WO 55.754, WO 44.677
11 WO 44.677, WO 44.133
12 Blackmore *op cit* 205, WO 44.132, PP 1849 ix
12 WO 55.754 29 Ja, 18 My, 3 Au, 21 S 1840, Blackmore *op cit.* 211
14 Blackmore *op cit* 212-6, WO 44.677.81
15 Rickword, G.O. *The West Essex Yeomanry Calvalry*, 1830-77. Blackmore *op cit* 275 226, PP 1849 ix
16 Blackmore *op cit.* 233-4, WO 44.644, pp1849 ix
17 WO 44.701
18 Blackmore *op cit* 232, WO 55.755, PP 1854 xviii
19 PP 1854 xviii 458, J. Stevens *Victorian Voices* 173
20 PP 1854 xviii
21 *ibid*, C 5116 1887

CHAPTER 3
Mass Production

Authorization was given on 18 February 1854 to purchase machinery for 'a new small-arms manufactory at Enfield'. Colonel Burn, the assistant inspector of artillery at Woolwich, was required to obtain from Joseph Whitworth details of gun factories and the makers of machinery in the United States. It was essential to the war effort that there should be no delay. The speedy completion of the new factory was regarded as a matter of the greatest importance to the public interest. Colonel Burn, with Lieutenant Warlow and John Anderson, were sent to America with instructions to bring back a model, or a 'specimen' (ie specification) of every machine used in that country for making muskets.

Lt-Col Burn arrived by Cunard in Boston on 10 March 1854; Warlow and Anderson, delayed by having to appear before the Commons committee, followed on 26 April. An itinerary had been arranged for them by Sir Joseph Whitworth, Col Colt and others. They had been authorized to purchase small-arms machinery to the value of £10,000. Their first visit, to the Government Arsenal at Springfield Massachusetts, found Col Ripley, the superintendent there, eagerly co-operative. The party was first interested in machinery for the manufacture of gunstocks. It was therefore necessary to check that the American machinery could cope with the close-grained English walnut, considered in England to be superior to the American variety. The machines worked well and Col Ripley recommended purchase from Ames & Co at Chicopee. The services of Thomas Buckland, the Springfield Armoury engineer, were offered and accepted.

From Springfield the three-man committee proceeded by train to Harford to visit Sharpe's Armoury which belonged to Robbins and Lawrence. Sharpe's breech-loading rifle and carbine were made there. It was a new factory, not yet completed but, said the committee, already 'conducted on the best manufacturing principles'. They were struck by 'the beauty and efficiency' of the machinery made at the works of Robbins and Lawrence. Mr Lawrence, who had accompanied them over the works, examined the parts of the 1853 pattern musket and offered to tender for the necessary machinery.

The committee then travelled on by railway to Newhaven to see Whitney's Armoury where the United States Rifle was being made. While there they visited a number of other factories: Jerome's, where 250 men were employed making

600 clocks a day at a cost of $1 each, Mallory's, where 2,000 locks and padlocks were produced daily, and a firm employing 175 hands in making 2,000 pairs of indiarubber shoes every day. At New York the committee were impressed by the mass-production methods employed in a further range of factories.

The estimate from Robbins and Lawrence was now received. The number of machines which, by this time, were seen to be essential made it obvious that the £10,000 allocation was inadequate. The number of stocking machines from Ames, for instance, would have to be increased because they turned the work out more slowly than other machinery and a bottleneck in the stocking department could not be allowed to hold up production. A modest increase to £12,500 was requested, but the Board, obviously impressed by the enthusiasm of their reports and depressed by the worsening situation in the supply of small-arms for the war, authorized the committee to purchase whatever machinery they considered necessary regardless of cost.

Their investigations continued in Philadelphia where they visited the Navy Yard, a factory making pistols by hand, a chandelier factory and Horstmann's Fringe and Accoutrement Works. Their curiosity and interest ranged widely over methods of mechanized production.

Remington's establishment in Washington made rifles for the United States Government. Remington was the first to introduce service barrels bored out of a solid bar of steel, straight through, using a vertical boring machine. All the steel had been imported from England.

The US Government Armoury at Harpers Ferry comprised two separate factories, one making the musket, the other the rifle. The machinery was similar to that used at Springfield but the factory was less compact.

At Pittsburgh the committee visited the Arsenal, three miles out of town, saw a rolling mill which had shut down because of the very hot weather, a biscuit factory, two iron works, a spade and shovel factory, wagon-building and plough-making works, a factory making nails, and another making iron spikes. They also examined the factory production of buffalo whips and melodeons.

Labour saving methods were essential in the US, said the committee, because of the high cost of labour. In consequence a wide variety of trades were carried on there in a way similar to the cotton industry in England, that is in large factories with machinery applied to every possible process, and using extreme subdivision of labour. A considerable proportion of the special machines which had to be employed were designed and constructed within these factories. As regards the class of machinery usually employed by engineers and machine makers, these were, upon the whole, behind those used in England, but ' in the adaptation of special apparatus to single operations ... the Americans displayed a degree of ingenuity which English industrialists would do well to imitate'.

The gap between the two countries was especially wide in the manufacture of small-arms. The two national armouries, Springfield and Harpers Ferry, and the private establishments of Col Colt, Robbins and Lawrence, and Sharpe's, were all constructed on the interchangeable system, every part being made by machinery on the premises from raw material.

The machines used for woodworking were roughly constructed of wood and in no way comparable in stability and appearance to the highly finished iron machinery used in England, but had the committee ordered them to be made of iron the delay would have been considerable and the price materially increased. A number of these primitive machines had been displayed at the Great Exhibition 1851 but, wrote John Anderson looking back in 1867, 'the ridicule which at first they excited was quickly dispelled when they were set in operation and at the 1862 Exhibition the display of wooden machinery had been on a grand scale'.

The committee dealt with as few firms as possible in order to keep things simple. A series of eight tenders were received and relayed home, all included the cost of packaging and delivery. There were approximately four dollars to a pound.

1. Ames 15 machines for stocking.	$30,860
2. Robbins and Lawrence 61 machines.	$17,660
3. Robbins and Lawrence 78 machines	$23,585
4. Robbins and Lawrence (woodworking machinery) 9 machines	$5,121

19 There were 18 boring machines in the late 1870s with seven spindles to each, each cutter took out its own length

20 The rifling machines, wrote Robert Smiles in 1877, have a peculiarly beautiful action. There were 25 in use and one barrel at a time passed through each machine

5.	Whitney tools for tinwork	$623
6.	Ames jigs and gauges	$5,600
7.	Ames 7 machines	$10,370
8.	Robbins and Lawrence barrel machinery	$38,748
		$132,567

These intelligent and well informed visitors from England had not confined their attention to the machinery required to equip the Enfield factory, they had examined a wide range of American industry and they were impressed. Factories

21 There were five polishing machines in the late seventies. The barrel was gripped by elastic hollow pads, having an up and down motion. One man could look after four machines

there had been built with ample workshop space, not only to ensure room for future expansion, but to make possible the systematic arrangement of the various processes. They admired the adaptation of special tools to minute purposes which enabled American manufacturers to reduce costs. They reported that in many of the factories a department was set aside for the express purpose of constructing or adapting such special tools and contrivances. Innovation in machinery was much more common in America and so many heads and hands were energetically at work at it that unless the American example was soon followed in England, American manufacturers would become exporters and not only to foreign countries, but *even to England*. Should this transpire the blame must fall on English manufacturers for their want of energy and foresight in improving their machinery and adapting it to special purposes. English industry had every advantage over American and ought to be able to undersell their competition on this side of the Atlantic.

Anderson, Burn and Warlow seemed also concerned to find that America led this country in the care bestowed on the comfort of the factory worker. The

22 The mill head or basin with the 1850s buildings reflected in the water. The fascia of this building was refurbished between 1967-9 revealing patterned ornamental brickwork. The tower, found to be unsafe, was restored in 1969, the bell refurbished and the clock overhauled

wash-houses were clean and sufficient, and presses were available where the worker could keep a change of clothes. There was abundant provision of good drinking water (that would scarcely have been available at Enfield Lock in 1854) in many cases with ice (the weather at the time was very hot) and the workers were less addicted to beer. Work discipline was admirable, there was little absenteeism and the men continued work up to the last moment of working time. They were on piece work wherever possible and this maximized production and reduced the need for supervision. The committee recommended its widest possible use in the new Enfield factory.[1]

Plans for the new factory at Enfield retained the canal, the basin and the water-mill alongside which steam-engines now provided additional power transmitted to the machines by belts and overhead shafting. Anderson's conception of a large room was also included. Anderson had planned a room 500 feet long and 180 feet wide with a 'saw' roof supported on iron pillars which also carried the shafting and the steam pipes for warming the building. Difficulties arose in the winter 1854/5 because the road east of the railway was flooded to the depth of nearly three feet. A platform had to be constructed to give access to the factory. The Royal Engineers began work on the foundations in the spring of 1855, but the ground was soft to a great depth, which delayed construction to such an extent that machinery was arriving before the buildings were ready.

Meanwhile the English weapons contractors continued to fail on delivery and, of the 90,900 1853-pattern muskets ordered, 61,242 remained undelivered on 24 April 1855. In those last few months of the war, before the new Enfield factory came into production, the Ordnance Department was reduced to a frantic search around the Continent to procure arms. Carl Hoffman the banker offered to sell the British Government a small-arms factory at Herzberg in the Harz Mountains. He also offered a factory at Lautesberg where iron was mined nearby and charcoal was available from the local forests. He promised to put these establishments into a state of high efficiency. His offers were rejected because it was hoped that the Enfield factory would be in production before Hoffman's proposal could be carried into effect. Some 1853-pattern muskets were manufactured for the British Government in Liège, some in America and some in France. On 29 December 1854 Lord Hardinge wrote to Lord Raglan that 6,000 rifles were on their way to the Crimea (in two vessels for safety reasons) and that 16,000 more would be delivered at Balaclava as they became available, providing of course that the English contractors kept faith. Production began at the Enfield factory in July and Sebastopol fell on 11 September 1855.

On 21 July 1855 the establishment at Enfield assumed the proud title the Royal Small Arms Factory (RSAF). It is significant that the new word 'factory' at this time replaces the old word 'manufactory'. On that date notice had been given that, in preparation for the change to mass production, the engagements of all workmen employed at the factory would cease. The skill and experience required for hand processes had been much reduced. Work had to be fed into machines, removed and checked for accuracy, the workers were mostly on piecework. A new skill however lay in the making of jigs and gauges in the toolroom. Several of the old hands were granted pensions, others 'with long and faithful service' were given gratuities, some were found employment in London or Birmingham, the remainder were re-engaged on altered terms. A new system came into operation on 23 July. Working hours were to be the same as those in private industry; Monday to Friday 6am to 6pm, with half an hour for breakfast and an hour for dinner, Saturday 6½ hours making a 59 hour week. The gates were opened at 5.45am and any workman who entered after 6.15am lost half an hour. The gates were closed at 6.30am. Wages were still paid on Saturdays before dinner.[2]

It was at this time that a branch of the Amalgamated Society of Engineers was formed at Enfield Lock with 48 members mostly from the toolroom. Branch meetings were held from November 1855 until 1870 at the Red Lion Enfield Highway. It was an arrangement for mutual profit; benefiting not only the publican. The branch could book beds at the Red Lion for ASE members passing through with travelling cards in search of work. One member's complaint about a bed at the Red Lion was investigated by the branch, but the complaint was declared unjustified. On another occasion the landlord was voted 5s as recompense for 'a nuisance committed by a member in the bedrooms'. A great deal of conviction and courage was required to become a trade unionist in the mid-nineteenth century for they were a small minority in the factory. Membership cost a shilling a week (one and a half hours' pay) and in addition there were occasional levies of 6d or 1s for the union headquarters or other causes. Often the hat was passed round at meetings for sick or unemployed members and fines could be levied for misdemeanours either in the branch or in the workshops. Meetings in these early years began with a roll-call, the condition of sick members was reported, then a list was read and posted of those members whose contributions were in arrears. Discipline at meetings was rigidly maintained, respect for the chair was paramount, no conversation was tolerated and any member with the audacity to cross the floor with hat on head was subdued by an accusing silence.[3]

A total of £240,593 had been lavished on the establishment at Enfield from January 1854 to March 1857; this included £75,986 for buildings, £64,667 for machinery, £27,270 for stores, £4,716 for salaries and £67,952 for wages. Some 1,250 men and boys were employed. With their families they crowded into the houses along and off the Highway and Turkey Street. Sanitary conditions in the area were appalling. 'I never saw a place so bad off', wrote a correspondent to the General Board of Health in 1857, 'for it is nothing but open privies and cesspools and the sewage from the houses runs from the cesspools into the wells'. There was no supply of pure water. There had never been a sewer or a road drain between the Bell and the Woolpack bridge. The place, he said, was swamped with filth. The land was flat and open ditches, stinking of sewage, stood almost stagnant.

The new factory was finally completed in 1859, fully equipped with machinery for the mass production of the 1853 pattern. By this time production had built up to become the main source of military supply for this country, delivering five times as many weapons as all other sources put together. The gunmakers of Birmingham were thereby forced eventually to adopt mass production methods. They employed very little machinery until 1861 when the new Birmingham Small Arms Company ordered a hundred machine-tools from Greenwood and Batley and from 1866 the company was engaged on the conversion of 100,000 muzzle-loading rifles to breech-loaders.[4]

The immense importance of the new works at Enfield, both locally and nationally, can been seen by the reaction to a serious accident which occurred there on 7 July 1859 when the great engine was put out of operation because the shaft broke. Several hundred men had to be laid off for a fortnight. They managed to keep on five or six hundred others because the machinery had forwarded sufficient work to employ them on hand labour. To meet the emergency the management was able to procure several small engines from Woolwich. They also sent to Stratford, to the railway workshops, to borrow a thirty-ton engine. It took three days, and, on parts of the journey, twenty-four horses, to drag it to Enfield Lock, only to discover that the bridge over the Lee Navigation was not strong enough to bear the weight. It had to be taken back to Stratford. A new shaft had been fitted by the beginning of August and the works were quickly back in operation. The factory had no assembly line and the time taken for any particular job was governed by the speed of the worker, mostly they were on piece-work and the prices were finely calculated to a hundredth part of a penny. Around 6,000 tons of materials a year, mostly coal, were delivered by barge to the RSAF along the Lee Navigation. Very little was taken out that way, but the Royal Gunpowder Factory at Waltham Abbey depended exclusively on the Navigation both ways.[5]

In the years 1859 to 1864, 423,939 weapons were produced at Enfield including 365,779 rifled muskets of the 1853 pattern. Production costs ranged from 37s to 39s 8d which compared most favourably with prices charged by the gun trade in London and Birmingham where weapons cost on average 65s. Those purchased in Liège cost 56s 10d. Some of the foremen had been recruited in America: there was Burton of course, Oramel Clark, foreman in the stocking department, Caulnin foreman in the smithy and McGee who was later to become manager. By 1859 the Enfield Rifle was the pride of the parish. 'Enfield', said the local *Observer* (June 1859), 'has very great reason to be proud of its rifle in the construction of which a vast and exquisitely perfect series of machinery is daily at work'. The article speaks of the sightseers and of the enormous steam-engine 'which sets the apparently complicated machinery in motion'.[6] The large room in 1861 was driven by two 40hp steam-engines, the barrel mill by a 70hp steam-engine supplemented by the water-wheels.

23 A former RSAF employee, H. T. Larman, recalled in 1928 that as a boy he came regularly to the factory to bring his father's meals. He remembered the large room (Machine Shop 1) being built in 1856/7. In 1859, when he started work, that portion of the large room, to the right of the main doors, was used for the stocking shop and fenced off with iron railings (shown in the illustration)

The factory was thrown open to the public on Mondays and Thursdays. It remained somewhat inaccessible. One visitor described (in *Chambers Journal*) leaving a dreary looking station called the Ordnance Factory. Crossing the line he found himself in a muddy lane with miles of flat meadows stretching out on either side and separated from each other by four-foot wide ditches filled with murky water. He came at last to a wooden footpath which led down to the Lee Navigation which he followed along the tow-path to the bridge leading to the Ordnance Factory.

Robert Smiles wrote the most glowing account of the works. 'We have reason to believe', he enthused, 'that there is a consensus of opinion among foreign visitors that the Enfield Factory is the most perfect of its kind'. He went on to describe 'the pretty water basin, ornamental as well as useful, in the centre of the square, with its clean stone coping all round'. The workshops he found were 'as scrupulously clean as whitewash, sweeping and scrubbing can make them'. The Smiles family had long been associated with Royal Ordnance. One John Smiles, first employed in 1804, had been appointed principal viewer in 1815 and had retained that position in 1840. Avid to discover the secrets of mass production, experts came from all over the world. February 1862 saw a second visit from Captain Davila of the Spanish artillery. On a previous occasion he had made an examination of the whole process of manufacture, he had taken copious notes and had endeavoured to entice away the chief engineer, Burton, the man who had come over from America to oversee the setting up of the new factory. Burton had been tempted to accept employment in Spain at a salary of £1,000 a year. This time Davila at least sought permission to engage skilled workmen in the Spanish

24 & 25 The large room, seen here deserted following closure, was planned by Anderson to measure 500 by 180 feet. The saw roof was supported on iron pillars which orginally carried the shafting and the steam pipes which warmed the building

service but Colonel Manly Dixon protested strongly and urged the Secretary of State to forbid such foreign interference. The Italians, the Sardinians and the Russians all requested detailed information, including ground plans, elevations and dimensions. In July came a visit from General Guyon, the Marquis of Andigue, and other officers of the French service, and in the following January the Japanese ambassador visited Enfield. The St Etienne chamber of commerce asked only for a drawing of the proof-house, while the Turkish army delegation requested a complete estimate of costs, with descriptions of the machinery, to facilitate the building of a factory at Constantinople where they proposed to produce 25,000 rifles a year. Visits by gentlemen cadets were in such demand that parties had to be restricted to ten. The RSAF proudly displayed its achievements at the 1862 International Exhibition, arms and parts were put on show with photographs of the machines and three specially commissioned drawings (31). RSAF pensioners were employed to man the exhibits.[7]

'Of late years', wrote Colonel Manly Dixon in February 1862, 'a system has been persued of having periodical inspections of arms in the hands of troops and the Volunteer Corps'. Only senior viewers were so employed. Col Dixon urged that the 6s a night allowance paid to them was inadequate, pointing out that it scarcely covered their expenses; 1s 6d bed, 1s breakfast, 1s 6d dinner, 1s tea and 1s for servants.

The War Office, in May 1862, acquired 42 acres adjoining the factory to construct a rifle range at a cost, including the land, butts, markers and huts, of £6,930. The formation in many parts of the country of volunteer rifle companies boosted at this time the demand for the 1853 pattern rifle. The 41st Middlesex (Enfield Lock) Rifle volunteers was recruited from factory employees in 1860. The uniform, paid for by the men themselves, was 'rifle' green with red facings, with shako and pom-pom. The corps had a brass band and a drum and fife band. It reached its maximum strength of 733 in 1868 and was disbanded in 1884. The *Enfield Observer*, in January 1862, reported that the factory was very busy, a great deal of overtime was being worked and several hundred men and boys had lately been taken on. All this brought a vast influx of money into the parish, wages paid out at the RSAF in the year ending 31 March 1863 amounted to £136,193, salaries to £3,150. An advertisement in the *Observer*, directed at the well-paid workmen, announced:

> Six-roomed houses, 1-5 Typo Cottages, Nichol Road (now Jasper Close) three beds, parlour, kitchen and wash-house, garden back and front, good supply of water.

Money could be found for good causes, £81 8s 1d was collected for the relief of suffering following an accident at Hartley's Colliery, and £220 for an organ in the factory church. With increased earnings on piece-work, income tax became a matter of concern to the Government. An assessment of the men's in-

26 The copying machine was extensively used in Enfield in 1859 particularly in making locks. One instrument moved round a pattern while another moved in exactly the same manner over the iron or wood to be cut

comes had been made in 1855 but the calculation of tax had proved too complicated for the War Office to deal with, for the artisans employed assistants and paid them out of their own wages, they also provided their own files, oil and other tools, and they had to pay for materials spoilt by negligence. Colonel Dixon insisted that such complicated adjudications should be left to the local commissioners of the Inland Revenue who had the power to call on the men for detailed returns and to require them to verify these upon oath. The Colonel moreover had serious doubts concerning the legality of the War Office making

27 Milling machines, or shaping machines as they were usually known in England, were the great feature at the RSAF

29 The smithy in 1877 had 86 fires, 20 steam hammers, eight screw drop hammers and three sets of rolls for making swords and bayonets

31 *These drawings (and that on page 66) were commissioned for the 1862 International Exhibition*

deductions from wages, especially if it was done weekly, for he held that the tax was only leviable annually.[8]

Conditions at the factory were now so improved that the men had time to enjoy recreational facilities. The manager petitioned the Secretary of State that five acres of the land recently purchased and incorporated with the factory should be made available for a cricket ground, quoits and a bowling green, to persuade 'the men to keep free of the public houses'. Temperance always provided a lofty argument for worthy causes in the nineteenth century. The dining halls at the factory were under the management of Truman and Hanbury, the brewer. Early in the year 1862 however, the buildings were taken over by the War Department. The brewer's tenants retained the exclusive rights to sell beer in the rooms, although no restriction was to be imposed on the men bringing in their own beer. The Royal Small Arms Tavern had opened in 1858, the Ordnance Arms by 1862. Temperance advocates faced a task as hopeless as that which defeated King Canute. Upon the ringing of the dinner bell, wrote one observer, the entire work-force surged out into the open air and headed for the public houses. They crammed into the bars and filled the benches outside, perched themselves on gates and rails to take their rough and ready meals, while the barmen and their assistants struggled to meet the demands for pots of 'arf and arf'. Time was short and the men were thirsty. But then, a quarter of an hour before they went back into work, came a pause, pipes were lit and they settled down for a well enjoyed smoke.

Sick leave with pay for up to two months in any one year could now be allowed

32 The Ordnance Arms opened in 1862

to men of good character with upward of three years service, this followed a circular of 25 November 1858. The medical officer at the time was Dr Savage, the dispenser Mr Steel. Foreman Lingard was one granted 14 days with pay, he had been employed since April 1840 and had had no leave of absence during all this time. Thomas Cockburn, one of the clerks, suffering a malignant swelling and ulceration of his leg, was given from 14 days up to a month to get treatment at Brighton Infirmary, he had had only seven days leave in eight years. Even alcoholism could be dealt with in a spirit of forbearance. Mr Munro was given three months leave for the recovery of his health, sanctioned, as Colonel Dixon earnestly emphasized, 'with the expectation that he will entirely abstain from the fearful habit of taking intoxicating liquors'. Medical equipment was also provided, for instance a truss for David Rosewarne who had ruptured himself at work.[9]

The management concerned itself with both the education and the religion of the employees and their families. A school and a church had been set up in 1846 within the factory compound. The school comprised one large schoolroom with two small rooms adjoining for a master and mistress. Control had been transferred to the National Society in 1855 at which time it had become a mixed school under a mistress with a staff of pupil teachers, it nevertheless continued to be paid for by the Board of Ordnance. There was a committee of management

33 Entrance gates were first erected on the east side of the bridge in November 1897. They were removed to the site shown here, west of the bridge, in September 1898 and a wall was built which took in the Girls and Infants School building, the front of which bore a stone with the initials VR and the date 1846

which, in June 1862, appointed Miss M.A.Cragg schoolmistress at £40 a year and Thomas Wadsworth as assistant master in the adult and evening school at £30 a year. A few months later he was replaced by W.Young at £40, the committee it was explained, 'being unable to secure the services of a person with suitable qualifications at a smaller rate of pay'. Expenses paid to the candidates by the Revd J.Harman, vicar of St James, on the occasion of the appointment, amounted to £20. In March 1878 the boys school was separated in preparation for a move into the old dining hall facing on the Lee Navigation, west of the factory church. The new schoolroom was opened by the Revd John Harman and a master was engaged with two pupil teachers, Charles Lynch and Bentley Baker. Matthew Arnold, poet and inspector of schools, reported on the progress of the boys school in October that year. He commended the premises, well-fitted with desks, and the discipline, but as to the instruction, the teachers being new, he made allowances.

Although it was a National school the factory management remained closely involved in the administration. There were two school managers, the vicar of St James and the superintendent at the works. The vicar was involved almost on a daily basis, taking classes in religious knowledge and intervening in minor matters. But the superintendent himself was so frequently called upon to settle disciplinary problems that it must have created an inconvenient distraction from his technical responsibilities. The suspension of the boy John Johnson for instance was referred in March 1876 to Col Close.

The school had the usual difficulties common to any working-class school. Boys played truant especially in late September to pick blackberries in Epping Forest, or to sport themselves at Waltham Fair. School numbers in the boys school constantly fluctuated between 80 and 120 according to the fortunes of the factory. Insubordination or irregular attendance could be punished by suspension, and pressure might then be brought on fathers working in the factory, although by 1871 the fathers of half the children attending had no connection with the works. Three brothers were suspended, for example, in October 1875, for bad behaviour, their father having encouraged them, it was said, to set school discipline at defiance. They were not readmitted until January and then at double fees. Short-time working in the factory often resulted in boys being kept away because parents could not afford to pay. The school continued to be supported by the War Office until a school board was at last elected in Enfield in 1894. Three years later the school was incorporated into Chesterfield, a newly built board school nearby.

Seventy worshippers had attended the factory church on the day of the Religious Census, 30 March 1851. The Methodists moved into the area and began worship in 1859, in a tent, obviously aiming to take the opportunity offered by the recent expansion of the RSAF work-force. They built a brick chapel in Grove Road that same year and looking hopefully to the future, purchased a large site in Ordnance Road in 1865. Not long afterwards a small Strict Baptist chapel opened in Putney Road and, in 1868, land was purchased in Totteridge Road for a larger General Baptist chapel which had become Strict Baptist by 1870.

A Mechanics Institute had been set up in October 1847 with a grant of £10 from the Board of Ordnance supplemented by a collection among the workers and a £3 donation from Robert Smiles. Income in the first year amounted to £45, it seems to have been spent on candles, snuffers and beer. Lectures and entertainments were held, in 1849 a slide entertainment was presented called 'Dissolving Views'. That same year a lecture on chemistry by Dr Letheby, lecturer on chemistry at the London Hospital, was taken so seriously that £3 was spent on apparatus. In 1878, the Institute having no balance in hand, a collection was made at the factory gates and £20 15s 6d was raised. George Grossmith the author of that delightful little book *Diary of a Nobody*, lectured there on 'Good Humour' and was paid 30s. Bookcases and bookshelves were provided in 1851 in a room appropriated for a library in the clerk's quarters. A considerable amount of money was spent on books, much of it on fiction for the lending library, Encyclopaedia Britannica was also purchased.

Dramatic entertainments were presented by the Sheridan Club established around 1866. Between 1892-1900 the curate, the Rev C. E. J. Carter helped to produce pantomimes in the Sheridan Theatre (also used as a dining hall). These

34 'The Milliner's Holiday' presented at the Sheridan Theatre

were performed by the workers with children from the school. Mrs Chambers recalled *Aladdin* when Mr James of Ferndale Road played Widow Twankey, also the *Sleeping Beauty* with Winnie Day, the station master's daughter, as the beauty, profits were used to improve the stage. The theatre fell out of use about 1908 and in 1920 the stage was dismantled and the properties sold for £61. The Institute provided £15 to start an orchestra in 1901 but an orchestral society was only set up in 1912. PSA (Pleasant Sunday Afternoon) concerts were given on alternate Sundays in the autumn and winter.

The Victoria Swimming Club was formed in 1887 at the Greyhound in Ordnance Road. Every August it held a race along the Lea from Enfield Lock to Waltham Abbey, hundreds of local boys were taught to swim. The scheme to provide a public baths at Enfield Lock was launched by an entertainment at the Sheridan Theatre and a committee was appointed, chaired by the superintendent at the factory J. Rigby. It aimed to raise £900; £650 had been collected by the time the Local Board had adopted the Baths and Wash-Houses Act in 1890 and Rigby appealed to the men, whose total earnings at this time amounted to nearly £6,000 a week, to raise the remaining £250.

Billiards could be played at the Institute on a table rented from the War Office in 1901, it was subsequently purchased for £15. So well used was it that a second table was later secured for £66. The men also played cards, draughts, chess and table tennis. A cricket club was launched in 1904, £8 having been spent to render the pitch in good shape and £20 on equipment. Both a rifle club and a tennis club were set up around the same time.[10]

Many of the men had recently moved to Enfield from Birmingham and elsewhere and were glad of any opportunity to visit family and friends back home. For this reason Col Dixon sought to secure the Monday after Christmas 1863 as an extra day's holiday. He pointed out that men returning from visits home would find it all but impossible to get back to Enfield by the Monday, indeed many men, over the previous Christmas holidays, had failed to get back on time and this had resulted in a waste of steam power on the Monday. The same was even more true of the Easter holiday which was the most important of the year for the working population. The Colonel in consequence sought permission to give the men an extra day following Easter, promising that the time so lost would be made up by extra exertion, or by working on Saturday afternoons. The sixteen viewers at Enfield, in August 1864, asked that they should be granted six days leave with pay yearly, these men had had no leave since the new factory began work in 1856.

There were, by 1862, eighty cottages for the workmen, some at 4s, some at 2s 4d a week, also cottages for the police sergeant, the schoolmistress and two bookkeepers. Nine larger houses were provided in Welch's Lane, now Ordnance Road, for Col Dixon and the management. The road west of the railway had been remetalled by the War Office in 1857 and in 1865 the 'parish' was paid £100 to widen the junction with the Hertford Road, but nearly twenty years later this work had not been done. Sanitary conditions in eastern Enfield remained poor. The dwellings along the High Road, including the lanes, yards and alleys leading off it, were crowded with humanity and the stench from the open ditches and from the filth on the sides of the highway was insufferable in hot weather. The state of the area was deteriorating, it was said, because the population had lately increased 'from the great addition made to the Ordnance Factory'. 'During the whole of last summer', complained Colonel Dixon in November 1862, it had been impossible to keep a window open at the front of his house in Welch's Lane in consequence of the 'disgusting and unwholesome effluvia arising from the holes in the road'. Now the winter had arrived, liquid sewage two or three feet deep lay in the ditches at the sides of the road. The Colonel urged the War Department to compel the Local Board to remedy the situation, but it was revealed on investigation that the repair of Welch's Lane was the responsibility of the War Office.[11]

Notes to Chapter 3

1 WO 44. 535, PP 1854/5 L 539
2 Putnam & Weinbren *op cit*, WO 44. 701
3 AEU Enfield Lock branch
4 PP 1856 xl, Pam, D. O. *Victorian Suburb*
5 Gazette Au 1959, SUPP 5. 872. 858, Putnam and Weinbren *op cit*
6 PP 1864 xxxv 595

7 Chambers J. 16 Ap 1859 p241, copy in Local History Collection, SUPP 5. 872, PP 1854/5 L 539-, WO 44.132
8 WO44.132, SUPP 5. 872
9 *ibid*, Chamber J. *loc cit*, SUPP 5. 872
10 ibid, Ms at Westcott, Pam, D. O. *Victorian Suburb*, Collicott, S. RSAF ms in Local History Collection
11 SUPP 5.872

Chapter 4
A Cloud of Uncertainty

The decision to equip the whole Army with breech-loading weapons was taken in July 1864 but the manufacture was delayed because the Committee on Breech-Loading Arms had not agreed upon the pattern to be adopted. In consequence of the uncertainty a hundred men had already been laid off at Enfield Lock and Col Dixon, anxious to avoid the loss of more skilled personnel, demanded of the War Office whether the staff reduction was to be considered permanent, or was likely to endure only until the department had decided upon and adopted the pattern for the new rifle. In this latter case, the superintendent pointed out, it would be ruinous to discharge skilled men who would prove most difficult to replace a few months later. The men, as well as the lads who worked with them, had nearly all learned their trades at the factory, it would require years to train replacements. Even when the new rifle had been adopted, alterations to machinery and tools would take at least four or five months. In the meantime the factory continued to produce muzzle-loading weapons.

The urgent need to update British weapons in the 1860s was the result of the change to the use of rifled barrels. This had created a problem in loading. It had been an easy matter to force a lead ball down the smooth bore of a muzzle-loading musket, but it was a much more difficult operation with a rifled barrel. Designers had therefore sought a method of loading at the breech end of the barrel and a committee had been established to examine the possibilities. Inevitably there was strong opposition to any change among the higher ranks in the British Army although the Prussians had used breech-loaders since 1848, indeed the employment of these weapons was to cause the collapse of the Austrians, armed with muzzle-loaders, at Sadowa in 1866. It was this very collapse, followed by that of the French, which at last forced the British Government to seek to emulate the Prussians.[1]

Dixon was opposed to the conversion of muzzle loaders. 'It will doubtless take as much time and expense to convert existing arms into breech loaders', he wrote in July 1864, 'as to make new arms, and the probabilities are against such converted arms being found efficient ....' Nevertheless the RSAF, by September, was experimenting with the conversion of the muzzle-loading Enfield rifle to the breech-loading Snider, which was said to give 30 per cent greater accuracy and an increased rapidity of firing in a ratio of five to one. The Snider action, invented by

Jacob Snider an American, could be easily fitted to the Enfield rifle, the breech block being hinged on the right-hand side at the rear of the barrel. At the same time the factory was investigating the use of steel for barrels and, in 1863, Dixon and Gunner travelled to Westphalia in Germany to examine the adaptability of puddled steel for this purpose. The German manufacturers were reluctant to allow them to see the whole process and Dixon emphasized the confidentiality of his report. Meanwhile English steel makers sent samples and, following tests, steel from Firth's was recommended. The Snider was the first rifle to be fitted at Enfield with a homogenous solid steel barrel. This had to be drilled with great accuracy through the centre of the rod, and every attempt to do this at Enfield had failed. The only place which had the necessary technical expertise was the Firth's Norfolk Works in Sheffield where the steel was obtained. A specification and an order for 2,000 barrels was sent there, they were to be conveyed to Enfield to be set and rifled with seven grooves.

At the time of the Fenian outrages in January 1868 an attack was threatened on the RSAF and on the Gunpowder Factory. Troops from the 49th Regiment were dispatched from Colchester Barracks to guard the buildings, and a number of employees were sworn in as special constables.[2]

The first ten years of the new factory had been a time of full and steady employment. There was a shortage of 1853 pattern rifles following the Crimean War and no private manufacturer was able to compete. From 1866 however severe disruptions were being caused by fluctuations and uncertainty in demand, for the Snider conversion was not altogether satisfactory. Lay-offs in July 1869 further reduced the numbers employed at Enfield. Several families were given free passage to emigrate to Canada and twenty artisans left for Constantinople, engaged by the Turkish government for two years to take charge of the production of Sniders. No provision existed for those thrown out of work, except the ignominy of the poor law.

In the year 1870-1 a supplementary grant of £7,500 was received to be employed to increase the production of 'any description of breech-loading rifle', and the money was used to convert machinery previously employed on the Snider to the production of the Martini Henry. The conversion of muzzle-loaders to breech-loaders, as in the Snider, had always been considered a temporary expedient. A more permanent solution had been sought and found in 1867 in a weapon using a breech action devised by M. Frederick Von Martini of Switzerland, with a rifled steel barrel with a small bore, .477 inches, designed by Alexander Henry of Edinburgh. The Martini Henry was officially adopted by the War Office in April 1871. The need to convert the machinery created a second and even greater upheaval at Enfield.[3]

It was at this juncture that Major General Dixon was given notice to retire after seventeen years service as superintendent, his successors were to be limited to a five-year tour of duty. Dixon's long experience was put to good use for he was appointed managing director of the National Arms and Ammunition Company in 1872 and subsequently became manager of the Birmingham Small Arms (BSA) Small Heath factory. Dixon was succeeded by Lt-Col Fraser. Many of the men were now left with little work for months and, during that time, when the machinery was being adapted or replaced, they were unable to earn more than a few shillings a week. Some found themselves forced to run up crippling debts in local shops. Toolmakers brought here from all over the country, were doing more than thirty hours a week overtime, paid at time and a half, but still the refitting progressed slowly. Work in the filing room came to a complete standstill at the beginning of January 1872 and the men, facing a prolonged lay-off, asked to be put on daywork to help with the retooling. The request was turned down because the Amalgamated Society of Engineers, most of whose members were employed in the toolroom, objected to gunmakers being allowed to work in the toolroom. The foreman supported the union and while men who had been employed in the factory for years were taking home starvation wages, more toolmakers were engaged from far-off parts of the country.

To accommodate the new machinery the smithery had to be enlarged and the former barrel-rolling room, no longer used since the steel barrels were now bored, converted to a grindery. The barrel-rifling shop also had to be extended and this caused difficulties because the vibration of the floor in the new extension pre-

35 Derbyshire grindstones, some of which weighed three tons, were used to grind the barrels. Grinding was an unhealthy trade and those employed at it usually died young

36 The gunmakers in 1872 were suffering from months of short-time working. A protest meeting in the Totteridge Road schoolhouse was addressed by Middlesex MP Lord George Hamilton

vented the cutting of the grooves in the barrels with the required precision. Further machinery for the manufacture of Martini Henrys was purchased in 1871-2 at a cost of £8,000. Five hundred machines had to be altered and adjusted and at the same time finishing machinery was substituted for hand labour for the manufacture of many parts.[4]

Much of the machinery was ready by early March 1872 but still the pieceworkers could earn no more than an average of 15s a week. Production remained at only 500 rifles weekly and many of these were condemned after inspection. Indeed no Martini Henry pattern rifle was issued to store until June. The gunmakers sought a meeting with the superintendent and although he himself could do nothing, he secured for the men the support of Lord George Hamilton, the Member of Parliament for Middlesex. Hamilton came down to Enfield to address a meeting in the Totteridge Road schoolhouse. The men told him that it would be another three months before they could earn their money, they had been idle for nearly eighteen months and many of them were deeply in debt. Inevitably they would have to seek further credit and if it were to be refused, then 'God only knows what would happen to the women and children'. The management had offered the men the opportunity to leave but this was out of the question, not only because of the cost of moving, but because they would thus forfeit any claim to superannuation or gratuities.[5]

The toolmakers at last departed. These strangers were remembered in Enfield for many years. According to correspondence in the *Observer* in January 1882, they had been the roughest elements from the worst districts of the Black Country... imported into the parish regardless of character or civilization, so highly paid that they 'drank their earnings from midday Saturday until Tuesday night and were constantly in trouble with the law'. The management had to issue a warning that any man whose name appeared in the papers, having been charged in a police court, would be immediately dismissed. The tooling completed they returned to Staffordshire and elsewhere.

The factory, by June, was so far behind schedule that night-workers had to be employed, they were paid the same piece-work rates as day-workers. The new rifle had a smaller bore and a flatter trajectory, which gave greater range than the Snider, and increased accuracy. The action body was made from a solid block of steel and a large cavity had to be formed to receive the breech block. Great difficulty was experienced in bringing machinery to bear to carry out this operation in mass production but this was overcome in November 1872. The solid bar was made hot in a furnace, then placed over a die of the required shape under one of Nasmyth's steam-hammers. This, falling with a force equal to five tons, worked night and day until 1,200 rifles a week were being completed. Two important improvements were made to the weapon at Enfield: an indicator to show when the rifle was cocked, and a safety-catch.[6]

The piece-work price for a Martini Henry had been reduced from £1 13s to £1 8s 9d and production reached 2,224 in the third week of January 1873, by this time a total of 32,092 had been sent to store. Even this rate of production however was far short of the 3,000 a week (without overtime) demanded by the Secretary of State. Estimates for the year 1872-3 for materials and wages amounted to £189,000 plus £3,000 for the purchase of further machinery to reduce hand labour.[7]

Working hours at the RSAF in 1855 had been fixed at 59 hours a week, before that time hours had varied according to the season of the year, averaging 55 hours a week. By 1872 the agitation among factory workers all over the country, commonly called the 'nine-hour movement', had been spread, through the efforts of the trade unions, to government factories. In July the superintendent at the RSAF received a memorial bearing 475 signatures demanding a reduction of hours to 51 a week, that is from 10 hours a day to nine, and six hours on Saturday, with a compensating increase of 10 per cent on piece-work prices. The claim was rejected, but Secretary Cardwell at the War Office, feeling that it was essential to make some concession, offered to reduce the hours to 54 a week, private trade having already conceded this. It could most conveniently be done by allowing the men to begin work later on Mondays; in this way it would cost the management little for it had long been notorious that many men absented themselves on Monday mornings after the revels of the weekend. Moreover on Monday mornings the fires had to be lit and the machinery set going, thus production was inevitably limited. The men had asked for time off at the end of the day, 'to enjoy their evenings at home, or in study', they said.

The men had holidays with pay, sick leave and medical attendance. The Secretary of State now extended the holidays by adding the Saturday between Good Friday and Easter Sunday. It remained at this time the expressed objective of the War Office to maintain conditions at the factory at a level superior to those of other small-arms workers, in order to retain the loyalty of the men. Even the current agitation, it was pointed out, had been supported by a comparatively small proportion of the work-force.[8]

The terms under which pensions were granted at the RSAF had changed a number of times since the new works had been built. These terms were now not only a source of dissatisfaction among the workers, but a cause of inconvenience to the management. Before 1859 all the men had been able to qualify for pensions according to service, as shown in the table below:

Pensions before 1859

Service	*20-35 years*	*over 35 years*
artificers	£20pa	£24pa
labourers	£10pa	£15pa

The Superannuation Act passed that year however provided that only men classed as permanent Civil Servants were to be entitled to pensions and they were to receive one-sixtieth of their pay for each year of service. To qualify as a Civil Servant it was necessary to pass an examination. The situation changed again only two years later, on 29-August 1861, when the War Office ruled that no man receiving the full market rate of wages would henceforth be entitled to a pension. Thus no man appointed after that date was pensionable, while their fellow workers, appointed before that date, and receiving the same wages, remained entitled to pensions. Problems were also created for management because a man's pension was calculated according to his rating and this could not readily be reduced. Piece-workers' prices therefore had to be regulated to allow a man to earn at least his rating although in many cases the introduction of machinery to carry out the process had obviated the need for skilled labour and the work could be done by a semi-skilled mechanic. Out of a total of 8,254 employed by the Ordnance in 1861, 4,585 on the establishment had been entitled to pensions; but by 1887 11,658 were employed while only 1,018 were on the establishment and were pensionable. It was granted that any man discharged after completing seven years service should receive a gratuity.[9]

The men were busy again and earning good money by the end of the year 1872 though most of them remained heavily in debt. That August the workmen had decided to make another attempt to set up a retail co-operative. The *Enfield Observer* was scathing, demanding to know 'was this the way these people expressed their gratitude to Enfield tradesmen who had so recently stood between them and starvation?'. One may look in vain through the local newspaper to discover what motivated the men to take this action so soon after the end of the crisis. Perhaps the shopkeepers had taken advantage of their customers' indebtedness to raise prices. The *Observer* continued to sneer, declaring that 'Past experience teaches the tradesmen that they have little to fear ... three previous attempts to set up co-operatives have proved miserable failures and large sums remain due to those who supplied them and to those who took shares'. The writer urged that the shop-keepers should learn a lesson, and in future they should refuse to grant credit then, the writer adds, perhaps telling the truth without meaning to, 'they could supply goods over the counter for cash at as cheap a rate as any co-operative store'.[10]

The idea of a co-operative had been revived at the suggestion of the factory management. The men had found themselves in such straights after all the short-time working that they had applied to Major Dyer, the assistant supervisor, for an advance of pay. They complained to him of the high prices which they were forced to pay in their local shops. The Major replied that then they must be fools to shop there; they should combine and with their joint capital set up a co-operative, 'similar to those achieving so much for the people in the north of England'. Twelve years earlier, when a co-operative bread society had been formed at Enfield Lock, it had failed disastrously. Nevertheless they now went ahead and founded the Enfield Lock Co-operative Association. Until such time as they could afford to purchase for themselves, they accepted the offer of the New Supply Association to provide the stock.

Shares were issued valued at 10s each and no member was allowed to own more than ten, they were to be paid for by weekly instalments of not less than 3d; thus the Co-op could serve also as a savings bank. The first purchase consisted of a chest of tea which arrived just before Christmas 1872. It was held at the secretary's house in Bell Lane. Members pledged themselves to buy one pound each to give the business a good send-off. Further purchases were made; a cask of paraffin oil, two cwts. of soap, six dozen of ' Price's snufless dips' (candles) and a further half chest of tea. All these were articles for which they had been overcharged in local shops. Business could be transacted only in the evenings for it was all done at the secretary's house. Cocoa, coffee, sugar, rice and tapioca were added to the stock. Sales improved when the secretary moved house to a better position on the corner of Bell Lane and the Highway. The Co-op however almost came to an untimely end in 1873 when stocktaking showed that stock to the value of £18 was missing. The secretary resigned and moved away.[11]

The skilled workmen were now (1873) learning how to voice their own grievances. For some years the cashier had acted as income tax collector, deducting tax from wages whenever he found workmen in receipt of more than a hundred pounds a year. The men considered this to be an injustice. Recent experience had shown that their earnings were subject to wide fluctuations from year to year and, although they might earn £100 one year, the following year it might only be £50. They believed that it was illegal to stop tax out of pay and between them raised sufficient money to seek advice. Counsel's opinion gave support to the feeling of the men; he declared that, since their earnings were casual, they ought to be assessed over three years. The men therefore instituted proceedings against Mr Clapham, the cashier, in the county court, but before the case was heard the Board of Inland Revenue caused the money to be returned.[12]

The Martini Henry was being produced at the rate of 3,000 a week by the end of the year 1873 and 40,000 bayonets had been converted to fit them. The modernization of machinery however had enabled management to dispense with much skilled labour, some men were discharged, some were pensioned off. A cloud of uncertainty hung perpetually over the prospects of the work-force. Orders fell off and by April 1874 were so low that a crisis meeting of local property owners was held at the Bell. It was rumoured that the requirement for 1874-5 was for only 673 pieces a week. The Martini Henry was not so popular as had been hoped, it was heavy and the recoil was violent. Declining prospects led to a fall in the value of property in the area. It was a situation unheard of near a government factory which had cost more than half a million. This situation had arisen because the Government was purchasing arms from private contractors, also from Germany. The men demanded that the railway company should provide workmen's trains to Enfield Lock like those to Liverpool Street from Enfield Town and Edmonton; the cheap fares, it was hoped, might attract new industry into the area.[13]

The mass production of a Mk I Martini Henry, with the weight reduced from 4.22kg to 3.97kg and the recoil diminished, was finally approved on 17 July 1874. The year 1874-5 proved better than had been forecast. More orders came in for new weapons and some 140,000 Martini Henry rifles were returned for modification, in fact there had never been so much work at Enfield Lock. The men were eager to earn money to pay off their debts, but management reduced the prices for piece-work by so much, the workmen complained, that they could barely earn a living and could put nothing aside for the future. They accused the management of replacing skilled men by semi-skilled labour merely in order to reduce wages, nevertheless a thousand extra hands had been taken on.[14]

The process of finishing by machine, using unskilled cheap labour, had again reduced piece-work costs from £1 8s to £1 3s 3d a rifle and the machine-finished parts, the management said, were more uniform than could be hoped for with hand labour, most of the parts could be assembled without adjustment. A year earlier 101 men had been engaged filing and altering parts at 2s a gun, now only 11 men were so employed at 9d a gun; filing was completely dispensed with in 1874. The large room, in which the filers had worked, now lay empty and was converted to house barrel machinery. The management were working on a plan for turning and finishing barrels, eliminating altogether 'the pernicious process of grinding', which hitherto had been the only important process carried on by hand labour. It was claimed that the RSAF would be the first successfully to achieve this end. Grinding was an unhealthy trade and those engaged in it usually died young. It was said that a grinder scarcely ever lived to be more than forty. Moreover the turned barrels would be more true and there would be a saving of money. The new system was in operation in 1874 and by the end of that year 65,000 barrels had been made by this method, accurate as they had never been before, and comparatively cheap, and grinding had been eliminated. James Thorpe described the factory, half a mile from the station 'along Armoury Lane', in 1876. He enthused about the 'series of large workrooms filled with machinery of the highest class and much of it of exceeding beauty and refinement'. The *Enfield Observer* waxed lyrical over the brilliant illumination provided by thousands of gas burners glowing late into the night, but the heat and fumes which these generated, (incandescent mantles were not introduced until the 1890s) with the lack of ventilation, made the factory unpleasant and unhealthy to work in.[15]

A new building was being erected for rolling barrels, with boilers, furnaces and an engine. By rolling the barrels the management hoped to save 3s to 3s 6d, or half the cost per barrel (but see below). The system of finishing by machinery was now completely installed, the machines had cost only £1,400 and there had been a saving, in the current year alone, of £38,578. The expansion of the toolroom had been completed by January 1875, ready for the installation of new machinery. A new steam-engine was being erected. Piece-work costs on the Martini Henry were further reduced from £1 3s 3d to 18s 9d, then to 16s 0d. The price of the weapon was reduced to £3 4s 6d.[16]

A great effort was being made to complete the year's order by 1 April 1875 with artisans working until nine on weekdays and six on Saturdays. The reduced prices paid for their work, the men argued, meant that when overtime came to an end they would be on starvation wages. Another grievance was the rigidity of the management's code of discipline. Although they were piece-workers they were locked out if they were a few minutes late. The rules had not been imposed so harshly when Colonel Dixon had been in charge, they said, he had instructed the police to apply the rules leniently.[17]

Fluctuations continued to plague the workers. The value of orders had fallen away by May 1876. The factory again awaited a decision on an alteration in the pattern. Great numbers of artisans and labourers were discharged. Many found their gratuities to be less than they had expected because, in calculating their service, the Government had made deductions for time lost, through no fault of the men, during alterations and replacements of machinery, also when there had been a shortage of work. The men had at other times put in many hours overtime, even on occasions working all night, but this was never taken into consideration. In one case, because a man had been sick for three weeks seventeen years previously, the Treasury had deducted a year's service.[18]

The Mk II Martini Henry was approved 25 April 1877 and 260,000 Mk I rifles were returned for alterations. New browning plant was installed for which a new building had to be provided. Piece-work prices were again reduced to 14s 4d and the cost of the rifle to £2 8s 7d. The factory was also engaged on the inspection of Gatling guns. The year 1878 saw the opening of the new barrel-rolling plant but subsequently doubts were cast on the worth of the process after it was discovered that barrels made this way were wearing out within ten years. A new method of drilling the solid barrel came successfully into operation, thus the management was at last able to by-pass the monopoly of Firth's in Sheffield. Stock-smoothing machinery was erected during this year but it still had not come into use in January. Barrels were proved three times; after being bored, after rifling, and at long range when the rifle was finished. Two to three hundred barrels would be tested daily in the proof-house. The store of gun stocks was reported to be very low, and Enfield held the only stock of importance in the country. New supplies were obtainable from North Italy and Switzerland. It was urged that a sufficient number ought always to be held in hand to avoid reliance on artificial dessication without which it required at least five years to season walnut. So much importance did the Board of Ordnance attach to producing stocks in our own country that it had planted walnut groves at all its stations.[19]

The new pattern Mk III Martini Henry was sealed in 1879 and that year the factory was engaged in the manufacture of 30,000 at a price of £2 13s 10d. Col Frederick Close, superintendent from 1876 to 1880, pointed out that the contract price charged by private industry was £2 18s 9d which, on the total production for the year, showed a profit earned by the factory of £47,550. Private contractors on current contracts had been permitted to continue making the Mk II. A dynamo-electric machine, worked by steam, was installed that same year to provide current for nickel-plating.[20]

The workmen remained under threat of sudden dismissal and some of them either could not, or would not, save. One woman who was sued in May 1880 for a debt of £3 10s pleaded that her husband, who had been earning £2 a week at the factory had been discharged eight weeks earlier when work was slack. He was now away in Birmingham looking for a job and she had nothing coming in to feed the children. Wages at the factory certainly had been reduced. Twenty years ago, wrote one correspondent in the *Observer* (14 January 1882), obviously pleased about the change, men had earned in half a week more than many of them did now in three weeks. He said that in the old days men had 'worked Wednesday, Thursday and Friday like horses and had spent the money like asses from Saturday through to Tuesday, and their wives and families were left worse off than those of bricklayers' labourers. The artisans of today', he claimed, 'and the workmen of twenty years ago are as different as is the present government establishment from the old building that gave place to it .... No one now has any fear of living too close to Enfield Lock, nor do landlords as a rule refuse to accept factory hands as

37 Government Row in 1881 comprised 88 cottages all occupied by RSAF employees ranging from viewers, foremen and pattern keeper to labourers, stokers and the gas lighter. Number 74 was occupied by seven young policemen, all unmarried

tenants'. The present artisans he thought were an industrious and saving class of men. The Mechanics Institute at this time had 300 members. It had a library of 2,000 volumes with a reading room in which were displayed almost all the daily and weekly papers, three American scientific journals and the best of the English reviews and magazines. Science and art classes were held with about fifty members.[21]

The men who worked at the RSAF and their wives came from all parts of the United Kingdom. At the time of the 1881 census the streets within a mile or so of the factory were mostly occupied by small-arms workers and their families. Many of their houses still stand in 1998 although few look quite as they did when the census enumerator called in 1881. It is possible however, here and there, to identify the houses and say something of the families who lived in them at that time. A total of 992 RSAF workers and their wives were named as living around the factory. Of these 162 had been born in Enfield but this number included many grown-up sons who had followed their fathers into the works. A further 98 had originated in neighbouring parishes: Waltham Cross, Waltham Abbey, Edmonton, Tottenham, Southgate and Sewardstone. Thus no more than one quarter of these families had originated in parishes near the factory. Some 250 men and

38 The ferry in 1881 was run by Robert Hatford, aged 69 from Burton-upon-Trent, and William Negus the ferry boy, aged 15, who lodged with him

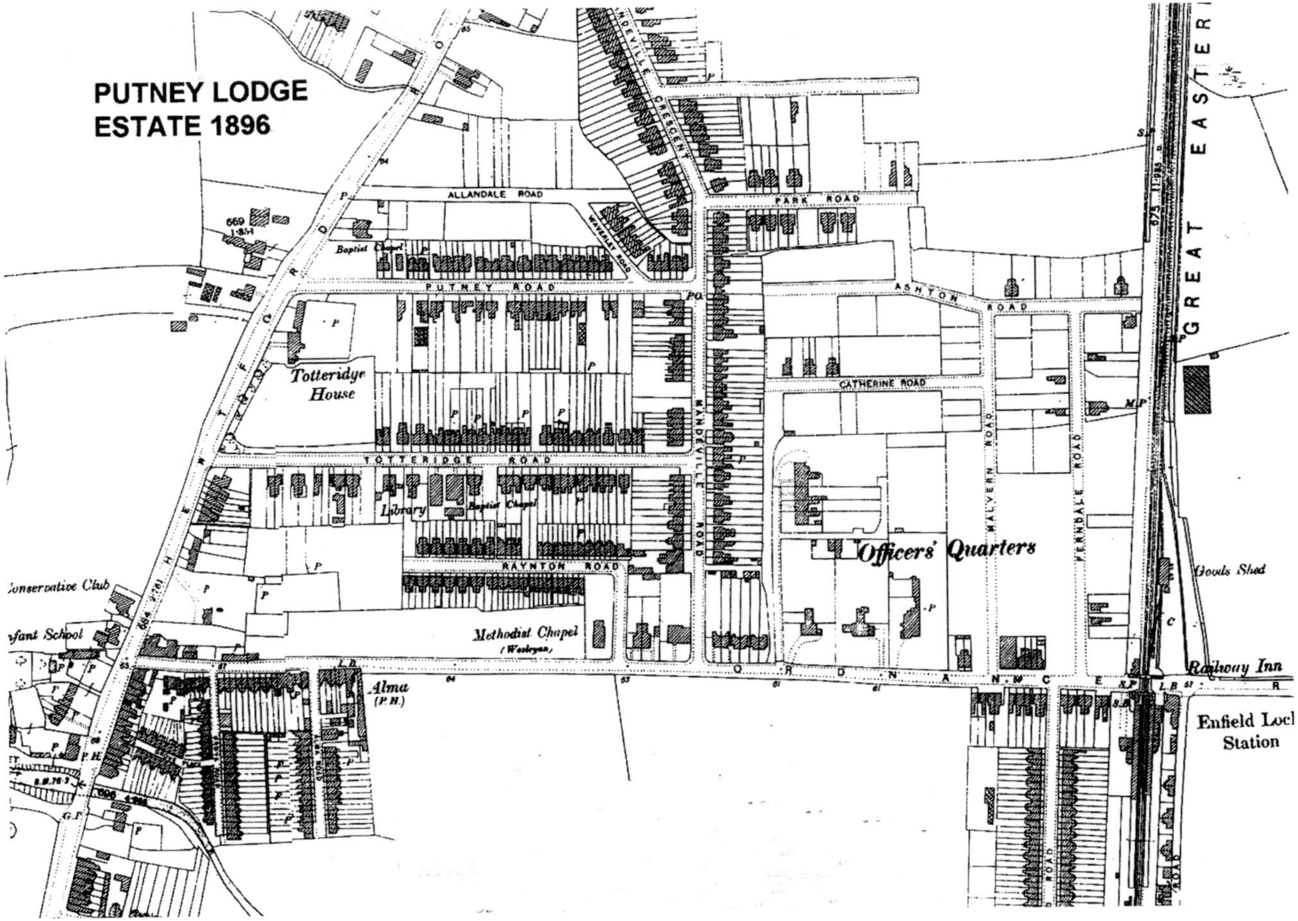

39 The Putney Lodge estate was sold to developers in 1867 and divided into 296 building plots for sale in Mandeville, Totteridge and Putney roads Most of the houses by 1881 were occupied by RSAF workers. The map shows the estate in 1896. (Reproduced from the 1896 Ordnance Survey map)

40 Lincoln Cottage, 3 Mandeville Road, a good detached house where lived, in the year 1881, Melbourne James Auber aged 48, a foreman at the RSAF. His wife, a year younger, had been born in Holbeach Lincs. James aged 16 worked as a warehouse-man at the factory. Richard 10, was still at school. Ten years later aged 58, Mr Auber had retired, James had moved away, perhaps married, and Richard was employed as a writer at the factory

women, a little over a quarter, had come from various parts of the London conurbation; from Bethnal Green, Whitechapel, Woolwich, Greenwich and Stratford. As might be expected 80 had originated in Birmingham, the home of the small-arms industry, and in the surrounding towns and villages: Darlaston, Dudley, Kidderminster, Sedgley, Willenhall, and above all in Wednesbury where no fewer than twenty of the workers and their wives had been born.

It is perhaps less to be expected that 70, about 7 per cent, should have originated in the West Country, not only from the larger places like Bath and Bristol, but from small towns and villages in Somerset like Pucklechurch, Selworth, Longport, Taunton, Wick, Kingsbury, Wells, Frome and Bedminster. They even came from as far as Devon: from Exeter, Tavistock, Budleigh, Southleigh and Paignton. There were Lockies and their wives born in Portsmouth, Portsea and Emsworth in Hampshire, and Whippingham on the Isle of Wight, from Devizes in Wiltshire and Fontwell in Dorset. Others came from Gloucestershire; from Bilson Green, Shipton, Cowley and Siston, places where none would have been aware of the small-arms factory at Enfield Lock. From Norfolk, Suffolk and Cambridgeshire they came: from Ipswich and Norwich, from Garboldisham, Aldeby,

41 Fern Cottages, Mandeville Road. The nearest was occupied in 1881 by Mary Stiles a widow (income from houses). She took lodgers, young men, John Harthorne a draper's assistant from Newcastle, Staffs, and his brother Samuel a millwright at the works, born in Liverpool, also Ernest Stainton an engine fitter at the works, a local man. In the further house dwelled Edwin West aged 56, an RSAF smith, born in Longport, Somerset, his wife aged 45 from Birmingham and two sons, both employed in the works, his daughter Emma, 13 was still at school. The lodger Walter Topple, an iron moulder from Lambeth, was employed at the RSAF. West and Topple were still working at the factory ten years later

King's Lynn, St Neots, Downham, Brancaster, Hepworth, Lakenheath, Kelsale, Haverhill and Withersfield. They had made their way from little towns and villages all over Essex and Hertfordshire; from Birdbrook, Maldon, Heybridge, White Roding, Finchingfield, perhaps as lovely then as it is now, from Stansted Mount Fitchet where the airport grows, from Matching, Sawbridgeworth, St Osyth, Great Bardfield, Thorpe Bay, Elsenham, Southend, Saffron Walden, Inworth, Nazeing, Farnham and Henham. In addition 24 were born in Ireland, 24 in Scotland, 33 came from Kent and 24 from Surrey and Sussex. This was a new and mobile population. A large proportion came from non-industrial places and the men would have had to learn the use of machinery and to adapt themselves to a factory regime.

There were industries in Enfield which might have been expected to provide work for the wives and grown-up daughters of the RSAF workers when times were bad at the Small-Arms. Ediswan's making electric light bulbs, employed many girls, it had replaced a jute factory at Ponders End in 1886. There was a

42 Totteridge Road. In the nearest dwelling, Myrtle Villa, in 1881, lived Richard Palmer a finisher at the factory, his wife Frances, aged 31, and his children: Ellen 7, Henry 6, Ferdinand 4 and George aged 2. Next door in Laurel Villa lived William Smith a gunstocker. The ages of his four children ranged from 7 to 14. Also with the family was his 80-year-old father-in-law described as an annuitant

43 John and Isabel Crocker, a gun-stock filer and his wife who lived at the nearest of May Cottages in Mandeville Road, were in their early thirties in 1881 and had four children. Ten years later they had lost two and produced three more, and his mother-in-law aged 77, had come to live with them In the further of the cottages in 1881 lived William Barber, his wife and eight children, in ages ranging from Albert 21, like his father a blacksmith at the works, to Lillian aged two

44 Eighty-five RSAF employees lived in Medcalf Road near the factory in 1881. Sixteen of these small cottages housed lodgers. George Clements and his wife with their growing family: Bertha, Ruth, Kate, Oswald (the only boy), Caroline and Lilly (7 months), managed somehow to find room for William Cobley, a fitter, with his wife and child. Most of the lodgers were young single men; with Widow Humphreys, for instance, lodged Arthur Aldridge, a sword-scabbard maker and Thomas Wilson, a machine fitter, both aged 24. Elizabeth Belcher a widow described herself as a lodging-house keeper, she had three lodgers, two armourer sergeants and an engine fitter born in Bombay. Medcalf Road was probably named after B. Medcalf who built Warwick Cottages before 1837

factory making black crape for funerals in South Street and the Cortecine, making floor cloth also sited at Ponders End. Joyce's hat manufactory at Waltham employed women as did the expanding market garden industry along the Lea valley.

The 1881 census shows however that few of their women worked. Out of 410 wives of workmen living near the factory, 395 had no employment. Many of course had large families of young children which gave them work enough. Only fifteen of the wives worked for a living. Many daughters over the age of fourteen must have left home, but of the 123 remaining with their families, over half (64) had no employment. Of the remaining 59, only four worked in the crape factory, and two in the jute works (still there in 1881) and four worked at the RSAF. Twenty-four young women were designated as dressmakers, needlewomen or milliners, nine were laundresses, eleven were domestic servants and eleven were teachers or pupil teachers and there was one governess.

There were no machine-guns in service in 1880 except Gatlings but that year a new machine-gun, the Nordenfeldt, had been introduced for naval service and a new section was added to the factory for the examination and repair of these. The parts for the Nordenfeldt were manufactured in Sweden, some were put together

45 In 2, Chester Terrace, Putney Road, lived James Duff aged 63 in 1881, he was a writer at the RSAF. Eliza his daughter was a teacher and her sister Marion a dressmaker, both had been born in Montreal over 30 years earlier

46 Alpha Cottages, Hertford Road. Three were occupied in 1881 by workers at the RSAF, William Ferrar a turner from Slapton in Devon, Henry Gwilliam a smith from Puddlechurch in Gloucestershire and Victor Rickmore a writer from Lakenheath in Suffolk

47 Warwick Cottages, Ordnance Road, looked out over the Lee Navigation. In number 8 lived Thomas Martin, in 1881 a 31-year-old viewer, with his wife Mary and their four children. In number 3 lived William Hubbard, a gun-barrel filer. His wife was a laundress, his 15-year-old son was a farm labourer, there were four other children, three of them at school and a 23-year-old lodger working as an agricultural labourer. The cottages were certainly in existence in 1837

48 Prospect Villa, Ingersoll Road, undoubtedly appeared more elegant in 1881. It was occupied by John Southin, a 48-year-old gunsmith and his wife aged 42. They had a large family: Evelyn, aged 20, was a dressmaker, Sydney 16 a pupil teacher and Charles 15 was an apprentice in the factory. The four youngest children were still at school

49 Maria Villas, two houses in Riley Road, were occupied in 1881 by the families of William Taunton, a RSAF fitter, and William May, a writer at the factory. Ten years later both families had moved on

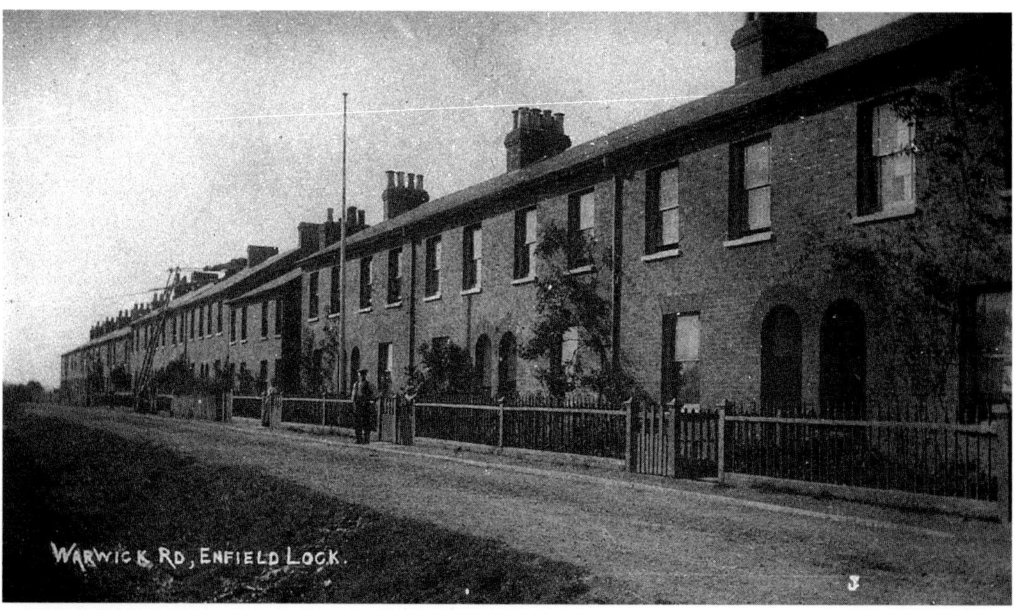

50 RSAF workers lived in 23 of the 26 houses in Warwick Road in 1881. They brought up large families in these little cottages. Joseph Axford, for instance at number 8, was aged 40 and his wife was 41. They had, so far, eight children, ranging from the 17-year-old Joseph, working like his father at the factory, down to the babies, Matilda, one-year-old, and Celia 13 weeks

51 Three houses in Officers' Quarters lay behind the larger ones on Ordnance Road. They housed the principal clerk and two junior clerks with their families

at the Henry Barrel factory in City Road and subsequently finished at Enfield, others were built at Enfield where a new pattern was devised by Col Arbuthnot for a five-barrelled gun to fire rifle ammunition. Col Arbuthnot also designed the breech block. The gun cost £120 at Enfield but £271 from Nordenfeldt. The following year (1881) saw the introduction of the Gardner machine-gun intended to supersede the Gatling. The Gardner Co was at Curtain Road in London. The Gardner machine-gun could be made at Enfield for between £80 and £90, although a royalty of £5,000 had had to be paid to the Gardner Co; no royalty had been paid to Nordenfeldt. Gardner spare parts were interchangeable and it fired .45 calibre rifle ammunition. The Gardner, like the Nordenfeldt, was originally designed for naval use and no suitable carriage was available when it had to be used ashore. The Nordenfeldt Company offered an experimental carriage to the Royal Ordnance and a contract for a number of these was allocated to the firm despite the opposition of Col Arbuthnot. When the carriages were delivered on board ship however the design was found to be capable of carrying only Nordenfeldt machine-guns, whereas by that time the Navy had 400 Gardners and only 20 Nordenfeldts. Col Arbuthnot reported on the matter with a justifiable degree of annoyance. The Gardner was recommended by the Ordnance Committee for general field service in 1885 but the number of machine-guns which could be made at Enfield was severely limited by the space available. Col Arbuthnot successfully pressed for a new machine-gun room and by 1887 the factory was capable of building four Nordenfeldt five-barrelled machine-guns a week, 1,500 machine-guns were by then in service. The Enfield factory expected soon to be dealing with Maxims.[22]

Col Henry T. Arbuthnot had been appointed superintendent in 1880. In his first year he had visited small-arms factories in the United States but concluded that Enfield had little to learn from the Americans. He had joined the Royal Artillery in June 1853 and had seen active service throughout the whole Crimean War. At the outbreak of the Mutiny he was sent to India, returning home promoted in August 1860. His appointment as assistant superintendent came in March 1862. The repairing department in Birmingham was established under his direction and he spent ten years there. The manager, James McGee, had also been appointed in 1880, after twelve years experience as assistant manager. He had been brought over from the Springfield factory in America where much of the machinery was the same as that at Enfield, where he had begun work in 1859 as a fitter and gauge-maker, by 1887 he was earning £400 a year.

The vacillating policies of Government which brought alternate rumours of short-time and threats of dismissal, followed by reports of substantial orders, continued to render life uncertain for the employees at the Royal Small Arms Factory. Although the decision in January 1885 to equip the Volunteers with 100,000 Martini Henry rifles, of which 80,000 were to be made at Enfield, had given promise of employment for five years, nevertheless only one year later (27 February 1886) came a report that 300 hands were to be discharged, despite the fact that an order had just been received for the manufacture of new arms for the regular Army. The management never found it easy to discharge men at Enfield, and skilled men had to be paid more than those at Sparkbrook, because mechanics preferred living in Birmingham or in any other large town.

The process of retooling continued to cause widespread disruption. In March 1886 a notice was pasted up in the works stating that it would be necessary to discharge 800 hands. The *Enfield Observer* (6 March 1886) expressed dismay.

> By the establishment of the Government Small Arms Factory, some thousands of the best and most skilled artisans have been attracted from the large manufacturing centres, and there has grown up what may almost be called a town of considerable importance in the district of the Highway. It was confidently believed that this neighbourhood would enjoy immunity from the prevailing depression. Only last year the Government determined to arm the whole volunteer force with the improved rifle, to be made at the Enfield factory. The work is not completed and will take five years at least. Many of the men, by frugality and industry, have been enabled to buy their own houses; others have become small shopkeepers. Many have their small savings invested in building societies. The 800 ordered to be discharged is but a slight proportion of those affected, for the families represent 3,000 to 4,000 souls.

The ASE had little power to protect skilled men from dismissal, indeed in 1886 five branch members were dismissed for taking a prominent part at a meeting to protest about the introduction of unskilled labour. Unskilled and semi-skilled workers had even less protection.

A delegation from the Enfield Highway Conservative Working Men's Club was dispatched immediately to Edmonton to see Lord Folkestone, the Conservative member for the constituency. He received them warmly and promised to raise the matter in Parliament. There were letters in the *Morning Post* and the *Daily News*, from Mr Littler of Broomfield Park and, by Friday morning, according to the *Daily News*, a telegram had been received at Enfield causing the superintendent to withdraw his notice. But uncertainty remained. Sir Henry Campbell Bannerman declared in the Commons that the order had been issued as a warning to the workmen of an impending reduction of wages, but he hoped that abrupt dismissals on a large scale could be avoided, nevertheless the rumours began again.[23]

Housing locally had improved. By the year 1886 the number of houses for sale in Enfield exceeded demand. 'It would be no great hardship', said the *Observer*, 'if the work of the speculative builder was suspended for a few years.' A number of the foremen, though very few workmen, had bought their own houses through building societies. Rents in Enfield were lower than those in London but the cost of living was higher. Some men from Birmingham and Sheffield came up to Enfield when trade was slack at home and went back as soon as trade there picked up. Despite the house building in eastern Enfield, many of the men, in 1887, lived three or four miles from the factory. The sons of old hands and men of the Army Reserve were given preference for employment; four hundred reserves, mostly unskilled or semi-skilled men, were working at the factory in 1887. Men over forty-five were not usually taken on.

The system of closing accounts on 31 March each year created difficulties for both men and management. In the early months of each financial year the smiths, doing preliminary work, were busy and everyone else idle, at the end of each year the situation was reversed. A workman might be left without wages for weeks; many provided for this, but the more improvident had to go in search of other work. The factory took the opportunity to do its stocktaking in the last three weeks of the financial year.

The *Sunday Times*, in February 1886, had complained that a quarter of the swords and bayonets supplied from Germany failed 'the bone and muscle test. Is it right', the paper demanded, 'to secure warlike supplies from abroad when Englishmen were prepared to take up contracts?'. An expansion of the Enfield factory was advocated so that there would be 'more English hands employed to do England's work'. The foreign swords so patriotically denigrated in the *Sunday Times* were purchased from Kirschbaum of Solingen, a thousand a month at 20s each, they employed the rolling process there. The firm not only kept meticulously to the terms of its contract, but the quality of the weapons had proved excellent. The only British firm which had been prepared to accept a contract for swords was

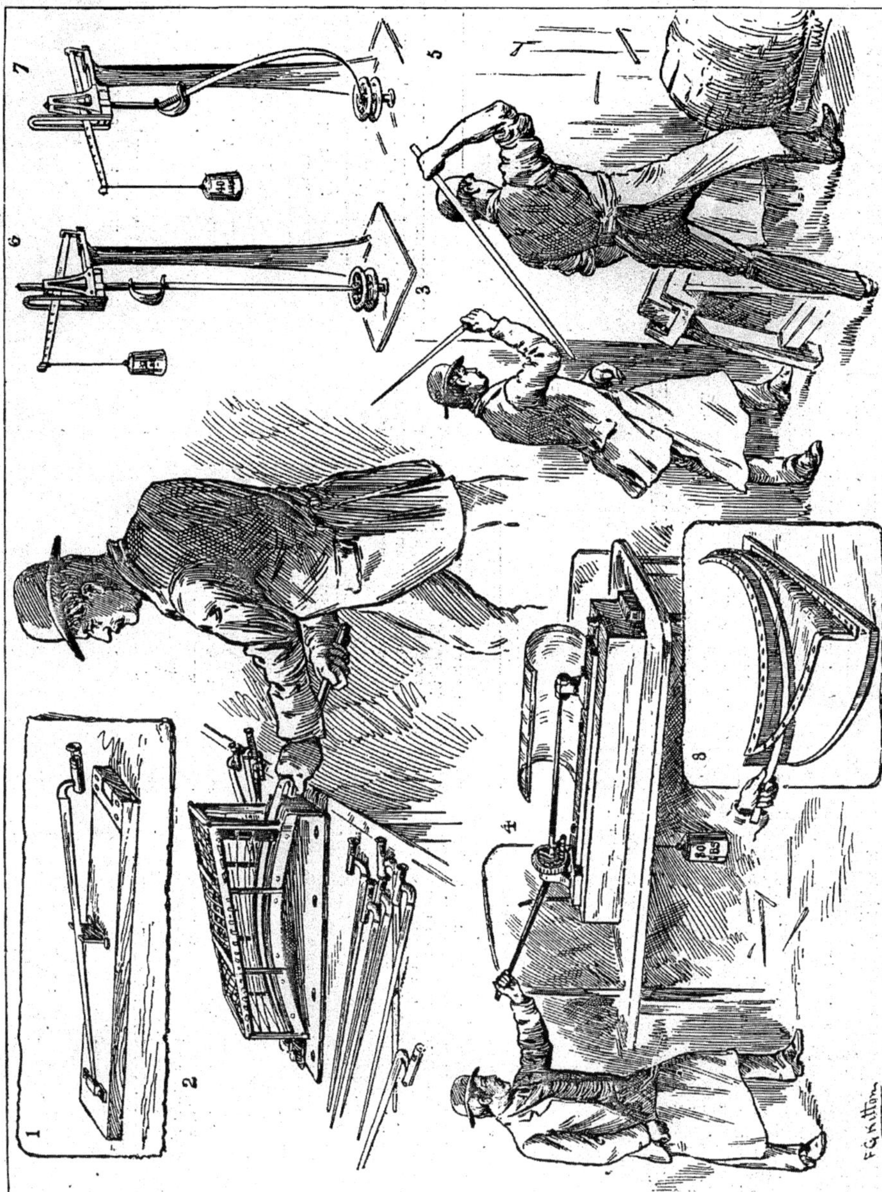

1. The "Bridge" Test (old style) for Triangular Bayonets
2. The "Curve" Test (the method adopted during the last two years)
3. The "Striking" Test for Bayonets
4. The "Twisting" Test for Bayonets
5. The "Striking" Test for Cavalry Sword Blades
6. The "Vertical Pressure" Test for Cavalry Swords (A weight of 32lbs. must not deflect the blade)
7. The "Vertical Pressure" Test. (A weight of 40lbs. must shorten the length of the blade by six inches without breaking it)
8. The "Curve" Test for Sword Blades

TESTING BAYONETS AND CAVALRY SWORDS AT THE ROYAL SMALL ARMS FACTORY, ENFIELD

Mole. The company had been given an order for 7,500 over three years, at 25s each, but although the War Office was paying them a higher price to encourage sword making in this country, the firm failed in both delivery and quality. A branch for the manufacture of cavalry swords had been set up in Enfield in 1885, some 3,000 had been successfully made at Enfield by 1887 with steel obtained from

Firth's of Sheffield. Rolling machinery at a cost of £400 was about to be installed at the RSAF where eight sword makers were employed. Col Arbuthnot, asked why he had found it necessary to place contracts abroad, confessed that he had been misled by the trade which had warned him that there was an essential secret in making swords, but he had since discovered that there was no such secret, and he proposed to give no further contracts to Kirschbaum. All bayonets for the Army had been made at Enfield for many years, none had been bought through the trade since 1878. New plant, being installed in 1887, would enable Enfield to turn out 3,500 rifles a week in normal working hours, with overtime the numbers could be increased to 4,000.[24]

A demand for small-arms had been maintained during the years following 1884 by the threat of war with Russia. Also the Indian government at that time made it known that it would require a large number of arms. Future prospects were enhanced by the decision of the War Office to adopt a magazine rifle. Private small-arms companies at this time appeared to be unwilling to increase their establishments unless they could be assured of orders sufficiently large to guarantee them against loss from whatever capital investment might be required and such a guarantee was unlikely in peacetime. Over the following year or two, few or no orders for small-arms were given to the trade.

Enfield at this time was capable of satisfying the peacetime demands of the British services. Foreign countries now had their own government small-arms factories; France had four, also one private firm, Austria had the largest small-arms factory in the world, and Russia had two factories. Thus there was little opportunity for British private manufacturers to export military weapons, and the orders received for these by private industry inevitably declined. Three private firms had equipped their factories for the mass production of small-arms but one of them, the National Arms and Ammunition Co., had to be wound up in 1882 after an attempt by the three companies to share government orders had been frustrated by the War Office. The Government had then purchased the company's premises which were at Sparkbrook, 108 men were employed there in 1887. All small-arms repairs were at this time carried out at Bagshot Street in Birmingham where 192 men were employed. The BSA had been hit by a shortage of work on arms in 1882 and had been forced into a premature attempt to manufacture bicycles.[25]

These circumstances had encouraged Col Arbuthnot, in 1885, to draw up a plan (53) for extending the RSAF in order to increase the production of rifles from 2,000 to 3,500 a week and to make provision for the manufacture of machine-guns and for magazine rifles. He was much circumscribed, he said, because of water on both sides of the factory buildings, but his plans were approved by the Surveyor General of Ordnance and others when they visited the works. A new store was proposed, a new smiths' shop (5 on plan), and a shop for the manufacture of machine-guns (4 on plan), all at an estimated cost of £38,000. A further

£12,000 was to be spent on the installation of three large dynamos from Messrs R. E. Crompton for lighting the works by arc lights. The generating station, supplying DC current, was to be sited on the east side of the canal (12 on plan). The superintendent had proposed tramways from the stores to the doors of each building, but this project had been deferred. Purchases of machinery by government small-arms factories amounted, in 1885, to £27,995, and in the following year to £21,852. This was by far the highest level reached in the whole period 1856 to 1900. At the RSAF alone £24,879 was expended in the year 1885/6. The *Gazette* reported in May 1886 that a large number of workmen were engaged on the new building, and that the new workshops would be fitted up with the latest patterns of machinery. A great deal of obsolete machinery had accumulated in the factory, much of it was sold in 1885 but only at the price of old iron.[26]

Col Arbuthnot was constantly on the look-out for new ideas, as he told the Inquiry of 1887: 'If I hear of machinery which I think will be of use, I usually go and look at it myself, then I send the foreman of the toolroom to give his technical opinion, and then buy one followed by others if it proves successful'. He cited the barrel-boring machine invented at Greenwood and Batley which had been first used in a factory fitted up by that firm at Terni for the Italian government. The superintendent had been impressed, he purchased one and subsequently six

53 The tumbling bay shown in plan 54 (opposite)

ARBUTHNOT'S PLAN 1885

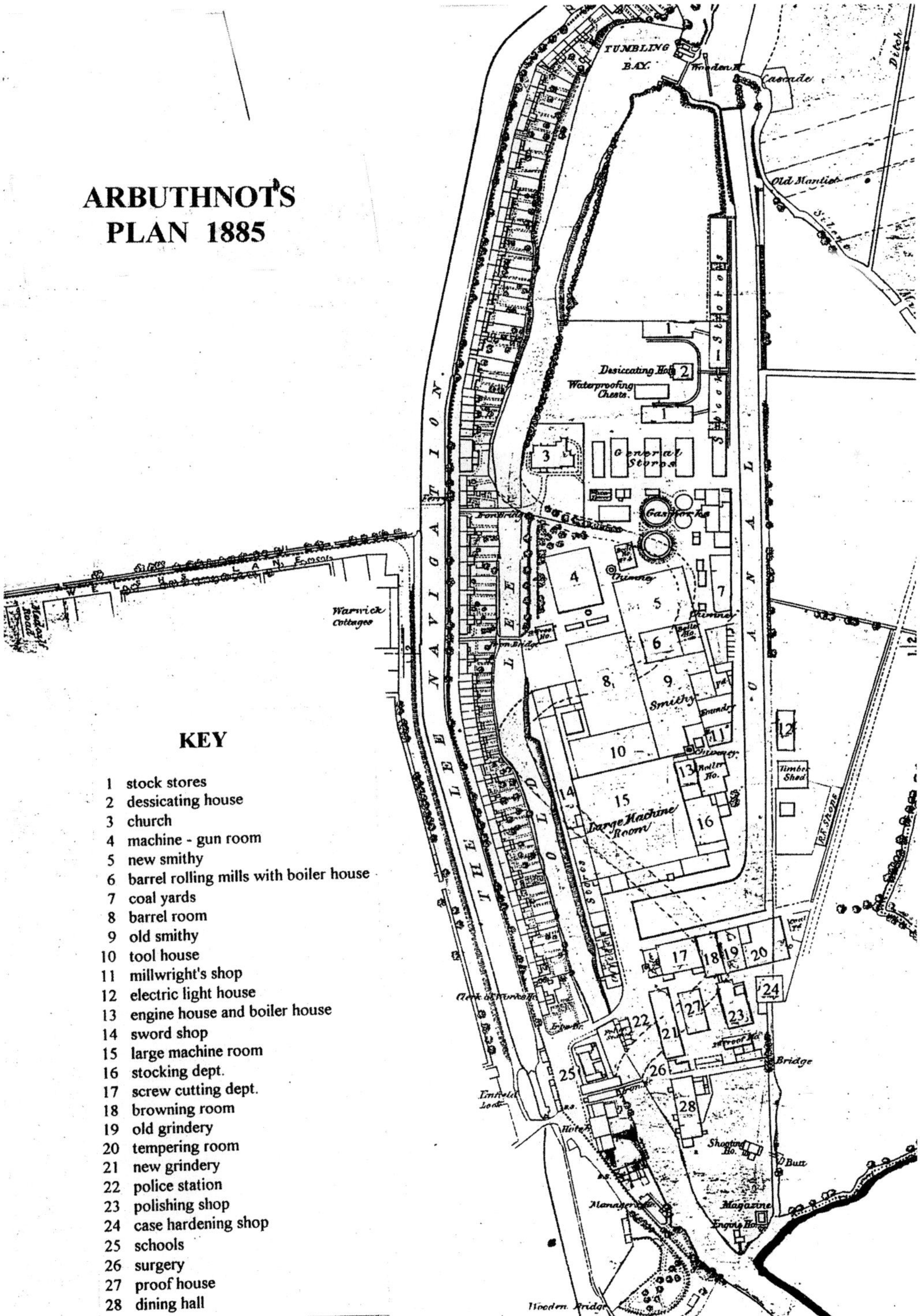

KEY

1 stock stores
2 dessicating house
3 church
4 machine - gun room
5 new smithy
6 barrel rolling mills with boiler house
7 coal yards
8 barrel room
9 old smithy
10 tool house
11 millwright's shop
12 electric light house
13 engine house and boiler house
14 sword shop
15 large machine room
16 stocking dept.
17 screw cutting dept.
18 browning room
19 old grindery
20 tempering room
21 new grindery
22 police station
23 polishing shop
24 case hardening shop
25 schools
26 surgery
27 proof house
28 dining hall

more. All the barrel boring at Enfield was now (1887) done by these machines with a saving of 1s 3d on each barrel, for, he said, 'one man can now do the work of six'. Arbuthnot, seeking new production methods, went over to Germany to see machinery being installed at Mauser's. He strongly advocated that the manager and the foremen should be allowed to patent their own improvements, at present the only way to reward exceptional talent was by rapid promotion; he instanced the foreman of the machine-gun room who had risen in seven years from the lowest grade of foreman to principal foreman on £5 5s a week. Colonel Arbuthnot had aquired considerable experience of magazine rifles, he had seen all produced in both Europe and America, over a hundred such weapons which, in view of the impending changes, should have been to his great advantage but he was replaced that same year by Colonel King-Harman.[27]

Militia battalions at this time had no armourer sergeant attached. Armourers, mostly military men, trained at Enfield, were sent to the units to examine and repair arms on the spot. Throughout the whole of the second half of the nineteenth century the small-arms industry had difficulty in coping with the rapid pace of innovation, production was delayed and the men lost wages while machinery was changed. According to the *Daily Telegraph* in June 1887, no fewer than eight structural changes had been made in the Enfield Martini rifle. At least 20,000, some said 70,000, of these weapons had been stored, yet every one of them would now have to be altered 'lock, stock and barrel'. The newspaper found it incredible that, 'in a practical nation like our own', weapons continued to be manufactured by tens of thousands only to be altered within a month or two. It was confidently predicted that, when General Smith's Small Arms Committee was able to agree, the Enfield Martini would be rendered obsolete by the introduction of a magazine rifle.The repeated alterations to the Enfield Martini, involving the remaking of forging blocks, cutters and other tools, necessitated periodic discharges and suspensions among the workmen. This state of affairs, the *Observer* predicted, would continue.[28]

In July a meeting was held in the Albany Hall chaired by J.C. Watts chairman of the Edmonton Radical Association. It was attended by some five hundred men who protested against further sweeping discharges. The Government, it was alleged, was to turn a thousand men adrift out of 2,400, many of whom had been induced to come from Birmingham on a promise of two or three years regular work; they would be faced with ruin. J. H. Matthews claimed that for many months these men, now destined to be discharged, had been doing overtime. This system, he went on, turned out 2,500 rifles one week and two or three hundred the next, and it was always the workmen who suffered. A letter from the meeting was sent to Lord Folkestone and questions were submitted in Parliament where it was emphasized that a whole community depended on the factory. Under pressure, Northcote, the Under Secretary for War, denied any decision to discharge, but his denial, it was pointed out by Matthews, was based upon a false distinction between suspension and discharge. 'If a man is put off with no wages it is the same thing', he asserted, 'these men are on piece-work and get paid according to the work turned out'. J. H. Matthews worked in the barrel room, his father Josh had formerly been employed in the factory and was president of the ASE branch in 1883. J. H. Matthews had taken a leading role in the Provident Society, acting for it as a health visitor. He was active in the Co-op and in the Enfield Highway Radical and Liberal Club. Nevertheless he was a landlord, with ten houses, and a moneylender.

The questions in Parliament and the publicity in the press, especially in the *Pall Mall Gazette*, delayed the discharges, but within two months the process was resumed. The Secretary for War, replying to Lord Folkestone, defended the dismissals, claiming that the number of hands was nearly double the number normally employed in peacetime. He blamed the long delays in settling the pattern of a new magazine rifle. Further questions were asked in Parliament in September 1887 in an endeavour to reveal the extent of the muddle. Had the manufacture of the Enfield Martini ceased, and if so when? Were those rifles, already made, being converted, and if so were there not already more than enough of them, since a magazine rifle with a smaller bore was about to be adopted? Enfield and Sparkbrook were turning out 4,000 Martini Enfield rifles a week in 1887, complete with bayonets, the steel coming from Firth's at Sheffield. It was suggested that the Enfield Martinis could be used in India following any replacement by a magazine rifle. The Committee on Magazine Rifles was aware of what was going on, for its members were visiting Enfield three times each week.[29] Yet the pace of these innovations had other consequences than the intermittent impoverishment of the work-force. Both the new arms and the conversion of the old ones created work for the factory. The repeated retooling of the factory could only be achieved by substantial capital investment. Innumerable improvements in machinery and in ways of employing them were initiated within the factory. These innovations eventually spread to the private arms industry and to the engineering industry generally.

The inspection of small-arms in 1888 became the responsibility of a separate and independent department set up in the factory but no longer part of it. Called the Small Arms Inspection Department it was under the Chief Inspector of Small Arms. Its duties were expanded to include all small-arms made by the Government, and those purchased from private manufacturers, the inspection of arms in the hands of the troops, technical weapons trials and the sealing of patterns used to guide manufacturers. Under its control were the Testing Section and the Pattern Room at Enfield with its comprehensive collection of small-arms. Thus the RSAF was placed on the same footing as other contractors.This control was to continue until 1989.

As a measure of self-protection a provident society had been set up in the works. Again the idea came from above, from Lord Folkestone when he spoke at the recently established Conservative Club in South Street. He suggested that the employees should put a small sum by each week which could be put out at interest. The scheme was initiated in detail by Col Arbuthnot, the superintendent, who became president *ex officio*. Mr Raitt the cashier at the factory became treasurer and remained so until his retirement in February 1900. The provident society was a purely private arrangement among the men, twelve of whom were elected to form a committee which met monthly. The rules had been drawn up by the men at a general meeting. Membership was not compulsory but 1,950 out of 2,200 belonged. Subscriptions were a shilling a quarter and the whole of the fund was expended each year, no balance was retained. Payment on death was £12, on the death of a wife £5. The society's income over the first three years up to March 1889 was £1,372 of which £1,272 had been paid out in benefits.[30]

Help in sickness was provided in the form of hospital and convalescent tickets for the men and their families, another arrangement set up by the men themselves. Men were not entitled to sick-pay until they had been in the department for three years, then they received half pay for a month, or for two months, from a sick fund to which the men contributed a penny a week. Workmen who fell sick could be seen by the factory doctor on any morning except Sunday, many of them belonged privately to sick clubs. Five days holiday a year were allowed; Queen Victoria's birthday, her coronation day, Good Friday, Christmas Day and an annual beanfeast.[31]

The daily routine in the factory remained highly regulated. Men coming into work at 7am (8am on Mondays) deposited a brass check in a box at the timekeeper's window. The gates were shut at seven and not again opened until two hours later. At 12 noon the men came out for dinner but, there being no facilities for buying food in or around the works, although there was a dining hall, the wives cooked meals for their menfolk which were collected by carmen and taken to the factory gates. Work resumed at 1pm when the men, going back in, deposited a copper check; the day finished at 6pm, it was then the duty of the chief workman in each shop to make an inspection and to report that all the lights and fires had been extinguished. When an insufficiency of work caused men to be temporarily laid off, the chief foreman issued each with a gate pass which had to be handed to the policeman at the gate.

Work was issued out by the chief foreman on work notes whereon was written the name and number of the workman, the work to be performed, the order to which it should be charged, and the viewer to whom it should be submitted for examination. The viewer, having signified his approval, would return the note to the foreman. Wages were calculated on the basis of these notes and were entered on each man's pay ticket which was issued to him before 12 noon on Saturday.

The limitation of overtime earnings for those paid by the hour to one-third of a man's full-time pay was abolished in 1882. The abolition created an incentive which ensured that the machinery would be used to its full capacity; the men could now earn 5s a week more and there was less need for supervision. The most skilled hands in the toolroom earned 1s an hour, 54s a week. In 1887 Mr Shepherd, the foreman there, had 262 men under him, working on gauges. Many of those who applied to work as first-class fitters were found to be insufficiently skilled. Boys were rated at 2½d an hour, they were apprenticed for five years and if proficient were kept on. Although piece-work was widely employed, every man had a daily rating entered against his name in the book, this was to govern sick and holiday pay. Piece-work prices were calculated to enable an average worker to earn one-third more than his daily rating; a man rated at 22s 6d for instance, might earn 30s on piece-work. If the men earned too much on a new job, management would cut the piece-work price, which was thus frequently revised, usually downward, until manufacture was in full swing, then it would tend to remain stable. The earnings of piece-workers doing the same work could vary by as much as 10s a week. Management at Enfield voiced no objection to trade union membership and in certain trades, such as toolmakers, fitters and blacksmiths, most of the men were members. Foremen were not permitted to be trade-unionists. Labourers were little unionized, many being new to factory work. Before being taken on they might have been shoemakers, bricklayers, bricklayers' labourers, or just working in the fields. They came in with the intention of earning what they could, about 22s a week as a day labourer, but the labourer might be put on a machine, a common milling machine perhaps, and then could take as much as 30s a week. Once he was proficient on the machine however he was no longer a labourer. The American arms industry used a system of sub-contractors, each employing forty to fifty men. The proprietor provided the machinery and the power, and the contractor found the labour. Management would contract with him to make perhaps 50,000 to 100,000 of a certain article. The system had been adopted at Enfield in about 1860. It was known there as the 'section system' and it enabled some of the foremen and others in charge of sections to take enormous wages, in some instances five or six times their ratings. And there was no proper check on their attendance. The system gave rise to many abuses, it was abolished by Lt-Col Fraser. Sub-contracting, a sweating system, was still used in 1887 at Bow and at Birmingham.

Reductions in costs could be achieved by the introduction of new machinery, or by the more economic employment of existing machines. Innovation in machinery was generally the result of co-operation. The new machine for drilling barrels horizontally, for instance, was a Belgian idea, further improved by Greenwood and Batley and perfected in Enfield. It had reduced the cost of drilling a barrel from 2s down to 3d. Any machine required for a special task would be designed and constructed at Enfield and if it worked, and more were required, a contract would be given to one of the specialist machine-tool makers. Only four or five firms in this country were capable of making gun machinery and the RSAF mostly ordered from only two, whose machinery required the least alteration.

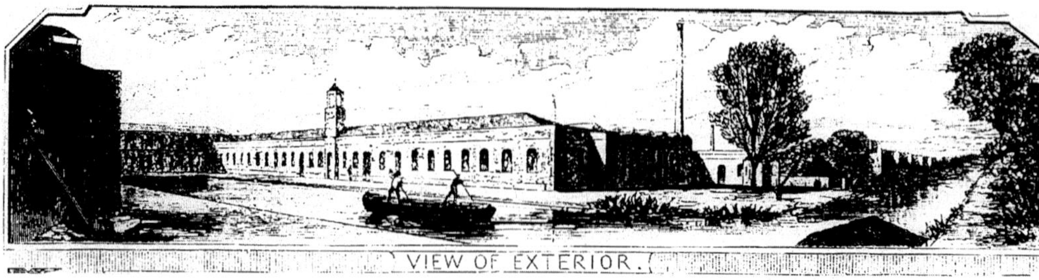

55 Coal, it was said in 1887, was brought in along the Lee Navigation in large barges belonging to lightermen as far as the Royal Gunpowder Factory. There enough of the load was transferred to smaller barges to enable the larger ones to negotiate the canal leading into the RSAF. This illustration of 1861 seems to show that the method was already used at that time

There was even collusion among the machine-tool manufacturers who by dubious arrangements sought to share the market, fix the prices, and even to equalize the time of delivery. Evidence of this is provided in an out-letter book belonging to the firm of James Archdale and Co of Birmingham involving William Muir and Co and Greenwood and Batley. Foremen who recommended or designed new machinery or improvements to weapons at the RSAF received no reward by entitlement, though they might get some increase of pay. In 1863 foreman Eaton brought to the attention of the superintendent an ingenious way of 'preserving the front sights'. It was fitted to 12 rifles which were sent to Hythe and Fleetwood for tests and reports. But there was no mention of any reward. In private firms such workers were given every facility and were rewarded liberally. A workshop existed within the RSAF for the construction of experimental machinery and making tools. A drawing office existed but with only two draughtsmen. In the first instance the innovating foreman had to explain his proposed change by means of rough drawings. Innovation was better facilitated in America where an inventor could secure a patent for as little as £12, patents were much more expensive in this country. McGee, the RSAF manager, thought that Enfield had the best machinery of any factory in Europe.[32]

Machinery could be more economically used by reducing the manpower required to attend it. There had been resistance from the unions when the management had required one worker to look after two machines (for planing machine-guns). The man objected and left, then they put two or three to work under the same conditions and they too objected. Finally the management insisted that the men must either work the two machines or leave. 'The men were sensible', it was said, they met and agreed. Although steam was the main source of power in 1887, water power was still employed. The one water-wheel which remained, provided 20hp. Coal, coke and wood were brought in along the Lee Navigation from the Thames, in large barges belonging to lightermen, up to the Royal Gunpowder Factory. There, enough of the load was transferred to smaller barges to enable the larger ones to negotiate the canal leading into the RSAF from the Navigation. In the year 1885/6, 5,308 tons of coal were used at the RSAF at a cost of 14s to 16s a ton. New boilers had been constructed within 15 yards of the coal wharf (6 and 13 on plan 53) and this had saved the labour used in wheeling the coal. Also delivered by barge, but also by rail, was crucible steel at £55 a ton, used for dies, sword blades and tools, and mild steel at £26 a ton for the bodies, as well as the tools which were sold to the workmen, such as files and hammers.[33]

Notes to Chapter 4

1 SUPP 5.872
2 *Gazette* My 1872, Rickword G. 'The West Essex Yeomanry' in *Essex Review* V.47 1948, SUPP 5. 872
3 *Gazette* Je 1869, Jl 1869, SUPP 6. 652
4 Putnam and Weinbren *Short history of the RSAF*, *Gazette* Mr, My 1872, SUPP 6. 652
5 *Gazette* My 1872
6 SUPP 6. 652, *Gazette* N 1872
7 SUPP 6. 652
8 SUPP 5. 1022
9 *ibid*
10 *Gazette* Au 1872
11 H.C.Smith *Co-operation in Enfield*
12 *Gazette* F 1873
13 SUPP 6. 652, *Gazette* Ap 1874
14 *Gazette* S 1874
15 SUPP 6. 652, *Evirons of London*, *Gazette* Oct 1874
16 SUPP 6. 652
17 *Gazette* Ja 1875
18 *Gazette* 13 My 1876
19 SUPP 6. 652, Smiles, R. 'The Royal Small Arms Factory' in *Great Industries of Great Britain*, 3 vols 1877-80 (BL 2268 e18)
20 SUPP 6. 652
21 *Gazette* 15 My 1880
22 SUPP 6. 652, C 5116. 360, 362, 358
23 *Gazette* 6, 20 Mr 1886
24 *Gazette* 20, 27 F 1886, C 5116. 359, 360
25 C 5116. 45, 50, 219, 355, 356, 360, *Gazette* 3 Je 1887
26 C 5116. 357, 527, Floud *British Machine Tool Industry* p.164, *Gazette* 15 My, 20 Au 1886, PP 1887 xiv 356
27 *Gazette* C 5116. 366-7
28 *Gazette* 3 Je 1887
29 *Gazette* 15 Jl, 19 Au, 2 S 1887, Putnam and Weinbren *Short history of the RSAF*
30 C 5116. 363, *Gazette* 26 N 1886, 15 Mr 1889
31 C 5116. 363, 367, 50
32 C 5116. 363, 50, 378, 382, Floud, Roderick *op cit* p. 178-9
33 C 5116. 378

Chapter 5

Boer War Prosperity

Enfield in 1887 was producing small-arms including 3,000 rifles a week, 40-50,000 bayonets a year together with swords and cutlasses. The factory had recently begun production of machine-guns not larger than rifle calibre. The management comprised the superintendent, the assistant superintendent and a captain inspector, all of whom were artillery officers, also a manager and a principal clerk. The work-force numbered 2,150 and included foremen, writers, men and boys. Also under the control of the superintendent were the 192 employees at Bagot Street, Birmingham. The Government had recently purchased the National Small Arms Company at Sparkbrook, also in Birmingham, but the factory was not yet in working order. Machinery was being erected there for the manufacture of carbines.[1]

During the last quarter of the nineteenth century rifle designers sought to achieve greater speed in loading. To this end advances had been made in the development of a weapon carrying a reserve of cartridges capable of being quickly loaded into the chamber from a magazine. The Germans were using the Mauser system, while in England the system invented by an American James P. Lee was favoured. Six hundred magazine rifles were to be tested by the end of April 1887, 300 manufactured in Enfield and 300 made in America under an American patent and of a different pattern. At Enfield the design had been worked out bit by bit, over four or five years, beginning with the Lee rifle which was a bolt rifle. Much work had been done on the magazine to see if it could be loaded on the rifle and that problem had been overcome with the help of one of the foremen. The weapon to be tested, which had been made at the RSAF, had fired 10,000 shots at Enfield and had never broken down. Another improvement being adopted on the Continent was a system of multiple loading by means of a clip holding five cartridges. The British Small Arms Committee, in December 1888, selected a .303 inch calibre weapon with the Lee magazine and action, and Metford rifling. It was known from August 1891 as the Lee Metford.

Rapid progress was being made as early as May 1889 in installing machinery for the manufacture of the new magazine rifle at Enfield and, by June, day and night shifts were being worked in some departments. Additional men were taken on and by September 3,200 were employed and 1,500 to 1,600 rifles were being turned out weekly. It was intended to expand production to 2,000 a week; £6,000

56 A rapid expansion in September 1889 of the number employed at the RSAF to 3,200, encouraged the brewer to enlarge the Greyhound

was paid out weekly in wages at this time. With the return of prosperity, militancy increased. A meeting outside the factory, seeking support for the London dockers' strike was addressed by James Rowlands, Member for Dartford, in September 1889. The men promised a weekly subscription and urged the dockers to go on to 'win a great victory for the working class'. The expansion of personnel at the RSAF persuaded the publicans to improve their premises, structural changes were made at the Greyhound, while the Railway Inn added a billiard room and club room. The factory church was enlarged to four hundred sittings.[2]

The new rifle was finally issued at the end of December 1889 to the Army, including the European infantry in India, but not to the native troops. It was thought at the time to be a perfect weapon. Said the *Enfield Observer* triumphantly, 'there is now a new rifle which has no hostile critics,' but after a moment's consideration, the paper added 'as yet'. It had been on the stocks for some years and was the fifth new rifle adopted for Army service since 1851. There had been the Enfield rifle from the beginning of the Crimean War, and this had been issued to the volunteers when these units were set up in 1859. The Enfield had been superseded in 1864 by the Snider which was the muzzle loading Enfield altered to a breech loader. The Martini breech loader had been adopted in 1871 but was not issued until about three years later, and it was not until 1877 that the whole Army had been equipped with it.[3]

The number employed at the factory reached a peak of 3,297 in 1890 as against

2,153 twelve months earlier. But the future looked uncertain for the wages estimate for the financial year 1891/2 was reduced by £75,000 and, by 1892, 900 men were discharged. The labourers were worst hit. Many of the skilled men who were laid off found work at the small-arms factory at Old Ford and the Great Eastern Railway ran a cheap train daily at 6a.m. to take them there, they were home soon after seven in the evening. But there followed more dismissals. A mass meeting, held in the market place at Enfield Lock in February 1892, was addressed by Henry Barrass who asserted that the Government was about to reduce production at Enfield from 2,000 down to 500 a week. The meeting protested that the work was being taken from Enfield and given to the gun trade. Barrass claimed that weapons could be produced much cheaper at Enfield, for example rifles at £3 8s as against £5 from private industry, Nordenfeldt guns at £125 as against £338, Maxim locks at £6 as against £20, and swords and bayonets 15 to 30 per cent cheaper. These lower costs he claimed were possible through greater efficiency, for wages at Enfield were higher than those in private trade where the system of sub-contracting or 'sweating' was still widely used.[4] Barrass came from Birmingham where local government was sophisticated. In 1897 he became the first Small Arms worker to be elected as a local councillor, in Edmonton. In 1903 he was appointed branch chairman of the Royal Small Arms Factory Employees Union.

A regulation of June 1892 compelled retirement at 65. Some 17 workers were affected ranging from Isaac Bowyer, a foreman aged 65 who received a pension of £160pa, to John Dixon a labourer, aged 69, who rceived £26pa. Ben Williams the timekeeper, nearly 73 years of age, was paid a gratuity of £33. Two labourers were paid gratuities of £23. The largest gratuity, £66, was paid to James Blanks an artificer.

The Birmingham Small Arms Co, by November 1891, had delivered 41,322 Mk I Lee Metford rifles and would be producing 840 a week by May 1892. The London Small Arms Co was producing 500 a week and would complete 30,000 by November. The companies, being informed that the price of the Mk I from Enfield was £3 12s, contended that such a price could not be believed.[5] Competition between the government factory and the trade was in some ways unreal. It must be remembered that the RSAF was never entirely commercial, free access was allowed to their competitors and models, drawings and gauges supplied. Thus private arms manufacturers were kept in touch with the latest improvements, and were enabled to copy machinery designed at Enfield.

By March 1892 output at Enfield had diminished by some thirty per cent. A deputation from the factory visited the House of Commons, questions were asked by Colonel Bowles, the matter was debated in Parliament and a visit by Members to Enfield Lock found two-thirds of the new and costly machinery standing idle. The MkII Lee Metford was approved 12 April 1892 and again the alteration of machinery and the preparation of new gauges at Enfield began. 'Personally', wrote

57 Enfield Lock railway station about 1900, looking east towards the factory. The engine, for those interested in such matters, is an M15. Many skilled men, discharged in 1892, found work in the London Small Arms factory at Old Ford, and the GER ran a cheap train at six o'clock in the morning, returning them home soon after seven at night

the Secretary of State, 'I should much prefer to get as many as possible of MkII from the trade'. Liberals in power were considered to be less sympathetic to the government factory than the Conservatives. Pattern rifles had been put in hand and the London and Birmingham Small Arms companies were asked to report how they could work to the new pattern. Both companies claimed that the price for the MkII, on a run of 50,000, should be £5, then they both compromised on £4 18s.[6]

The RSAF remained closed on Saturdays and the labourers, able to work only five days a week, found their money reduced from 19s to 17s 7d. The cessation of Saturday working had hit the labourers badly for they were paid by the hour. Henry Barrass wrote, in February 1893, to protest to Campbell Bannerman, the new Secretary of State for War, about the loss of jobs at Enfield; 'Has [*sic*] things are looking just about as bad at Enfield as they were in the Tory regime of last year, I think it is a most fitting time for you to see a copy of what the committee did,' and he enclosed a printed leaflet setting out the men's grievances. 'Men are being discharged every week', he went on, 'and many hundreds are on short time and if something is not done we must start the agitation over again'. The Secretary of State replied coldly that the discharges were imperative and in no sense peculiar to Enfield.[7] In answer to a question, Parliament was informed that the lowest wage paid at the RSAF during the previous week was 11s 1½d, to a lad of eighteen doing piece-work, but twenty-two men, over twenty years old, were earning less than 18s, the lowest being paid 16s 8d. 'We are informed', said the War Office spokesman, (Campbell Bannerman) 'that the current rate for labourers in the neighbourhood is 16s'. 'The farmers', he went on, 'have in fact complained that we are raising the rates in Enfield and Waltham'. This assertion was strongly denied by a correspondent in the *Observer* who obviously had firsthand experience of local conditions. He pointed out that the market gardeners in the area, Messrs Wilson senior, Wilson junior, Mitchell, Humphreys and others paid 18s to 20s a week. Over and above this their men had many opportunities considerably to increase their pay, such as planting, bunching and gathering, all of which tasks were done on piece-work, enabling the men to earn 30s a week. The current wage for a sewage worker in this area was 20s 3d and builders' labourers were paid 5d or 5½d an hour for day work. Campbell Bannerman moreover had made a comparison between a skilled labourer, working in unhealthy conditions in a factory, and an unskilled labourer working in the fields. The hardship caused by low wages was added to by ever increasing rents.[8]

The movement to secure sixpence an hour minimum pay for labourers at the factory, to equal what the dockers had won in 1889 and what had lately been conceded by the London County Council (LCC) and generally in the Metropolitan area, was set on foot by a meeting at the Greyhound in Ordnance Road in December 1892. The cost of living in the Metropolitan area, it was pointed out, was one-third lower than it was in Enfield. A deputation was appointed to seek the support of Col Henry Bowles of Forty Hall. The London Trades Council also lent its support and a committee was set up with J. H. Matthews as chairman; Henry Barrass was a member. Henry Barrass was much respected by his fellow workmen. Many of them gathered on 12 November 1896 in Mr Catlin's restaurant at the Woolpack bridge to present him with a gold chain and an illuminated address to honour his work in improving conditions on their behalf. The cuts had made the Liberal government unpopular in Enfield and the men at Enfield Lock remained staunchly Conservative. On the day of the general election in July 1892, when the works closed at midday, a long procession of carriages was laid on by the friends of Colonel Bowles to take the men to the polling booth in the main road where for an hour the polling was brisk. Unionist checkers confidently claimed that the voting was four to one in favour of Henry Ferryman Bowles the Unionist candidate. He was returned by a substantial majority. Another meeting was held early in February 1893 when Bowles shared the platform with Keir Hardie. It is quite remarkable that this occasion should have brought into such close proximity a Unionist landowner and a socialist ex-miner, by then the Independent Labour MP for South West Ham.

There were 155 labourers employed at the factory; most received between 4½d and 5½d an hour, only a few got sixpence, and some men with families received only 3d. Tom Mann, who had done so much to help the dockers win the 'tanner' in the 1889 strike, addressed a large open-air gathering outside the works in April 1893 and called for a 48 hour week. This was granted in all Ordnance factories in 1894. At Enfield the hours were, from that time:

Monday	7.30am-12 noon, 1pm-4.30pm
Tuesday to Friday	7.30am-12 noon, 1pm-5.30pm
Saturday	7.30am-11.30am.[9]

The sale of refreshments at the factory during working hours was stopped. There was no increase in piece-work prices at this time, but the day-workers did well, since their weekly pay remained the same as for a 54 hour week. The fortunes of the work-force continued to fluctuate. The Liberal government (1892-5) was accused of running down the Enfield factory. It had been reported in the *Evening News* in March 1893 that the management was about to discharge a further 300 men. 'The RSAF', said the paper, 'presents a woeful spectacle with hundreds of big machines lying idle'. The estimate for the wages for the year 1893/4 was £120,000 against £160,000 in the previous year. A valuation of the buildings and machinery at over £200,000 was given in the House of Commons. Some work was created when all the serviceable Martini Henry rifles, which had been exchanged for Lee Metford magazine rifles, were returned either for store or for conversion to carbines at Enfield. The government small-arms factory at Bagot Street Birmingham was closed in 1894 after a hundred years. Work was transferred to Sparkbrook and a large order for rifles was allocated there. After a while work at Enfield picked up temporarily and by July, six-day working had been resumed. Short-time working however recurred in March the following year and Major Bowles relentlessly resumed his questions to the Secretary of State for War. The uncertainty of employment held wages down, but the necessity for another change in rifle design was to predict early prosperity.[10]

The cordite now used in cartridges caused so much erosion in the breech end of Metford barrels that it rendered them unserviceable after only 4,200 rounds. An answer to this problem was found at the RSAF by the use of a barrel with deeper grooves in the rifling, which could be fitted to the Lee Metford. The new weapon was brought into service as the Lee Enfield MkI in November 1895. This brought about a revival in the fortunes of the factory and with it the opportunity for a pay claim. Those employed in the large room claimed that their rate of pay had not increased for twenty years. The men held a meeting in the canteen in March 1896 to demand an extra penny an hour; there was no strike threat, only the appointment of a three-man delegation consisting of Henry Barrass, J. Moxam and J. Adams, to see the manager, James McGee, who gave them a sympathetic hearing.[11]

By January 1898 conflict between the British and the Boer Republics of southern Africa was becoming inevitable and overtime was being worked at the Royal Small Arms Factory. The Boers were importing arms from Germany and engag-

58-61 Mafeking was relieved in May 1900 after a siege lasting seven months. The victory was celebrated with wild enthusiasm. In the factory, workshops were decorated and the people of Enfield, four thousand of them, made the journey down Ordnance Road to celebrate with the workers

ing German artillery officers. Prospects at Enfield had so improved by April that men were flooding into the district and there was a surge in house building. Said the *Observer* (22 April 1898) 'stacks of bricks and building materials are being carted into various open spaces'. An order for twenty thousand rifles was rumoured in May. Yet the optimism had been premature and by March the following year the Government, with its usual wisdom and foresight, was laying off men because red tape required that new orders could not come in before the beginning of the new financial year in April. In May, however, came an order for 114 machine-guns. Demand for rifles remained low although they could be turned out in Enfield at £2 12s against a cost of £3 4s in Birmingham. The editor of the *Enfield Observer* (6 October 1899) expressed his pleasure to find 'that the alarm-

Pretoria
Avenue

ing reports ... that a state of war had been precipitated by a Boer invasion of Natal is nowhere confirmed'. The rumours of war may have alarmed the editor, but they presaged prosperity for the artisans of Ordnance Road. The machine-gun shop began working overtime, as did the smith's shop and the toolroom, making parts for wagons and tent shoe pegs, but orders for rifles still did not come in. Collections were made in the works, as also at Ediswan's, to provide for the wives and families of fellow workmen recalled to the colours. The events of the South African War were given wide coverage in the local press. It was going badly and a series of reverses and surrenders in December 1899 brought gloom and despondency throughout the country. Seven marksmen from the Enfield Lock detachment of the 1st Battalion Middlesex Volunteers embarked for South Africa on 13 January 1900.[12]

Orders for the Lee Enfield at last began to come in with the new year. Some departments of the factory were working until midnight. Soon a night-shift had to be employed and Sunday working began. Further work was provided when imperfect sights on this rifle had to be replaced. It was the Middlesex unit which reported that, in the guerrilla warfare now being waged by the Boers against the British Army, the Lee Enfield could not compete at short ranges with the Mauser, supplied to the Boers by the Germans. The fault lay in the sights, and new backsights had to be hurriedly manufactured at the RSAF and at BSA and sent out to units on active service. The war engendered in the middle class a concern for national defence with the result that rifle clubs were set up throughout the country. Mr Devine, the schoolmaster, offered the range at Claysmore, his home in Clay Hill, for the use of the Enfield Rifle Club.[13]

The spring brought a series of British victories which culminated in May with the relief of Mafeking. Enfield rejoiced; flags and bunting were hung out, bonfires were lit and there were fireworks. The Foresters' band and the boys' band from Chase Farm School played patriotic music. The Royal Small Arms Factory was thrown open to the public, each department competing in decorating the workshops. Patriotic fervour brought thousands on a pilgrimage to Ordnance Road; on the first day nearly four thousand sightseers crossed the bridge. Through August 1900 the men were working until eight o'clock every night. The flight of Kruger into exile in the autumn seemed to presage the end of the war, but guerrilla fighting continued for a further eighteen months. It was a time of glorious prosperity at the RSAF. Peace was not established until April 1902, after which the prosperity faded.[14]

The wages estimate for the year 1902/3 was down £27,000 on the previous year, foreshadowing less overtime and less employment. The Lee Enfield had been criticized on account of its excessive weight and length, and the Small Arms Committee had been considering a newly designed weapon. Late in the year 1900 the superintendent of the RSAF had sent the committee a memorandum suggesting that time and expense would be saved by adapting the existing Lee Enfield, for it had many good qualities. The result was the Short Magazine Lee Enfield introduced into service on 23 December 1902. It was shorter and lighter than its predecessor, therefore easier to carry, it was robust and serviceable, it was to last, with minor changes, for more than fifty years; it saw this country through two world wars. As a consequence of this change an order for 25,000 Lee Enfields was cancelled at the beginning of 1903 and the War Office ordered the necessary retooling of the factory. Much of the machinery had to be adapted and new machines were laid down. Production was not resumed for many months during which time the workmen had to exist on reduced pay; the imposition of unreasonable conditions added to their grievances. J. H. Matthews became their self-appointed spokesman. At a meeting in the large hall at Chesterfield School, he shared the platform with James Branch the Liberal parliamentary candidate. Matthews told the piece-workers that they should be organized to resist the employers. He and Henry Barrass had been employed at the factory for nearly thirty years, he said, and formerly, when a man had no work on a Friday, the management had paid him his wages and let him go home. Now, men were forced to come in on Saturday mornings to get their pay. In former days workers were paid whenever they were unable to get into work because of floods; now if they were unable to get in, they got nothing. The following week a public meeting was held to protest against the rundown of the works, for the depression also hit the local shopkeepers and they were forced either to give credit, or to sell nothing. Many of the men had worked only one to three days a week since Christmas and they were coming out of the gates every Saturday with only 5s to 15s in their pockets. The Urban District Council dispatched a letter to the War Office telling of the extreme distress among piece-workers. J. H. Matthews called a meeting in the dinner-hour one day towards the end of September 1903, to urge the formation of a union branch of the Royal Small Arms Factory Employees Union. The branch was formed with Matthews as president and J. McGrath as secretary to represent the piece-workers and labourers. Matthews had played a leading role in every dispute at the factory since 1886 and, in 1907, he was selected by the United Workers Federation as a parliamentary candidate. It is strange to see that the author of a history of the ASE Enfield Lock branch, even in 1970, regarded the RSAF Employees Union 'as a forerunner of house unions', and although it became affiliated to the TUC, the local ASE branch called upon its own executive council to declare it 'unrecognised' and refused to negotiate for, or in co-operation with, its members.[15]

A reduction of the labour force from its wartime level looked inevitable yet, a year after the fighting stopped, 2,600 men were still employed at the factory against 2,100 before the war. The War Office was of the opinion that at least three or four hundred men would have to be discharged, but in the meantime hundreds were kept on though earning £1 a week or less. They hung on in the hope that the factory would return to full employment once the plant for the new rifle

62 In 1904, following the Boer War, the workers were on short time and the factory remained closed on Saturdays. Two-thirds of the machinery was standing idle, children were going to school ill-fed and poorly clad

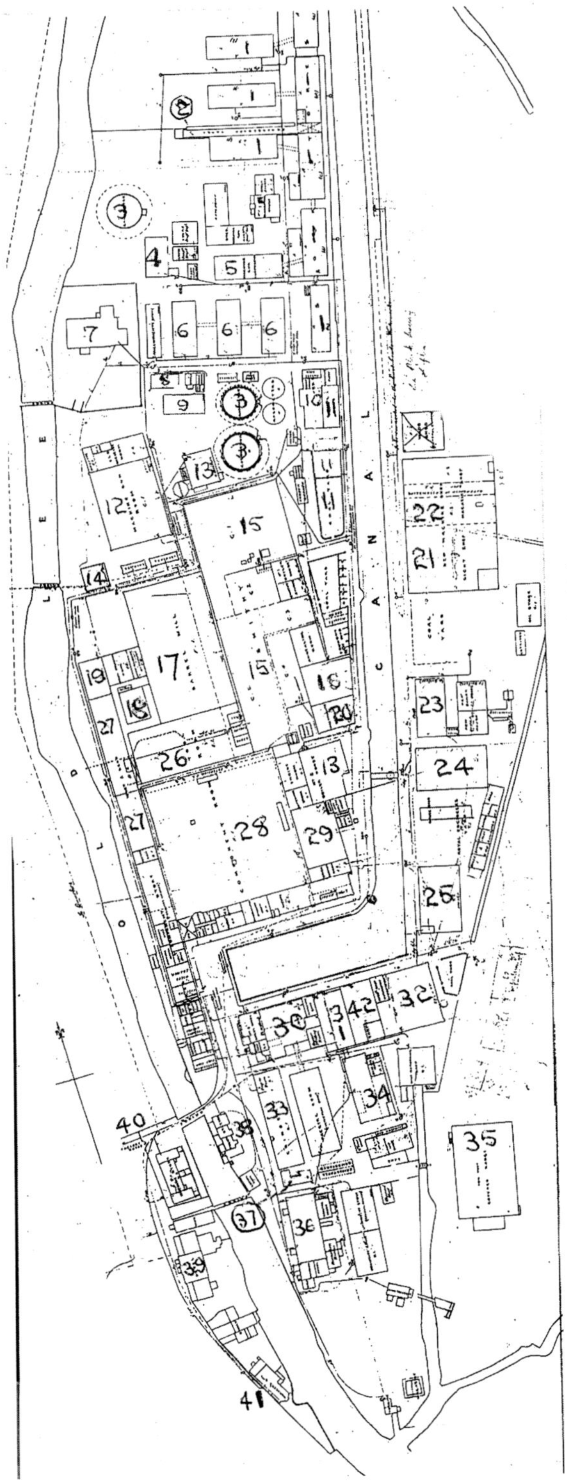

RSAF 1904

1 stock stores
2 chronograph range
3 gasometers
4 inspection dept
5 machine gun inspection
6 general stores
7 church
8 water tower
9 pattern room
10 gas works
11 coal yard and sheds
12 machine gun room
13 boiler house
14 view room
15 smithy
16 foundry
17 barrel mill
18 latrines
19 brazing room
20 millwrights
21 sight shop
22 magazine shop
23 electric light house
24 stock filers
25 new screw cutting
26 tool room
27 sword, bayonet, scabbard
28 component room
29 stock finishing
30 old screw cutting
31 browning
32 hardening
33 grindery
34 polishing
35 covered range
36 dining hall
37 surgery
38 police station
39 hotel
40 factory gates
41 manager
42 carpenters shop

had been laid down. A question in Parliament from Colonel Bowles, at the end of March 1904, revealed that 152 men had earned, on average, less than 22s a week over the previous six weeks. If they were family men, they must have been living in poverty. Many men had run into so much debt that when they were finally discharged their debts exceeded their gratuities, for which they often had to wait five or six weeks without a penny coming in. They complained that these gratuities were calculated in such a way as to rob them of their due; for while management deducted the time when men were laid-off because there was no work, it took no account of the overtime they had put in. Thus men who had worked the required seven years qualifying time were, by dubious adjustments, often robbed of their gratuities. J. H. Matthews led a deputation representing the United Government Workers Association to the War Office to voice these grievances. With the impoverishment of the work-force, houses were falling vacant as people crowded together to save on rent. House building ceased and the distress spread to the building trade. Much of the trouble was said to be caused by imperfections in the design of the new rifle; production could not start until solutions had been worked out.[16]

Short-time working continued throughout the year 1904; the works remained closed on Saturdays, reducing the hours to forty-four a week. The average wage for the week ending 3 December 1904 for the general run of skilled and semi-skilled workers (excluding labourers, foremen, leading hands and viewers) was £1 15s 6d. Four hundred men were discharged in March 1905. Seventy-five per cent of the shopkeepers in the Lock area, declared Councillor David Weston, were on the verge of bankruptcy. Men who had tried to purchase their houses through building societies had been unable to keep up repayments. Children were going to school ill-fed and poorly clad. Two -thirds of the machinery at Enfield Lock was standing idle. The number of empty houses in the area was shocking.

The Murray committee, appointed by Balfour in November 1905 to look into the working of government factories and workshops, published its report in 1907. It set a figure for the number of employees at Enfield at between 1,900 and 2,000 and limited the weekly wage bill to about £4,000. Further it revealed certain principles which were seriously to effect the RSAF and its workers. It was urged that in the allocation of government orders one-third should go to Ordnance factories, two-thirds to the trade. Ordnance factories should be permitted to expand only up to that point from where it would be possible rapidly to increase production in an emergency. The work assigned to the Ordnance factories must therefore be no larger than could be carried out without taking on extra staff and without working overtime. The committee reassured itself with the thought that, since the Ordnance factories had had no difficulty in securing extra labour during the South African War, they would have no difficulty in the event of any future outbreak.

Nevertheless it was taken into consideration that since the Ordnance factories held vast capital in the form of land, buildings and above all machinery, this capital ought to be employed in the most economic way possible. Therein lay the dilemma. The production of goods other than arms was considered but rejected as an interference with private enterprise. Reduction in numbers employed appearing inevitable, it was piously but unrealistically proposed that it should be achieved by natural wastage not discharge.

The United Government Workers Federation continued to demand that labourers' wages be increased from 22s 6d to 26s. This, claimed J. H. Matthews, the president, was the money paid by other local firms. He cited John Rochford, Ediswan's, the London Brick Company and others. The Royal Small Arms Factory remained short of work. Production was held up by doubts about the efficiency at long range, of the new short-barrel rifle. According to the *Daily News* the minister, Haldane, at the War Office, was being pressed in parliament to place more orders for bayonets at Enfield where they could be made for 8s 1¾d each, as against purchase from the trade at 9s 8d. Four hundred men were dismissed from the Waltham Abbey Gunpowder Factory between March 1905 and May 1906.[17]

The production at the RSAF of the Short Magazine Lee Enfield in the financial year 1906/7 amounted to only 47,000 but 19,000 rifles had been sent back for conversion. The busiest department was the barrel shop, making all the barrels for both new and converted rifles, 66,480 in all, as well as 14,000 as spare parts. The toolroom was busy making gauges for the new Mk III rifle, and also gauges for the Indian factory. The average number of men employed each week was 2,027, and the wages earned amounted to £193, 433. The maximum number of men who could be employed at Enfield in 1907 was 3,500 and the minimum number who had to be retained to enable this maximum to be reached without delay was 2,000. The original power house had been demolished and work had begun, in January 1907, on the installation of two additional 350 kW sets. During the disruption this caused, one of the turbo-generators had to be closed down and throughout the interval the load from the large room had to be transferred back to the beam engine. This was finally dismantled in 1910 and the engine house converted to the tinsmith's shop. The old plant for illuminating the factory by gas was virtually redundant after the introduction of mercury vapour lamps in 1910. The only additional shop converted to the use of electric power during the year was the machine-gun department, for no further load could be taken when the lighting was in use. The management looked forward to converting more of the shops to electricity when the two new generators came into operation.

The factory fire brigade had been reorganized in 1902, it now comprised one professional fireman, one warder and 32 trained volunteers, 20 of whom lived in Government Row from where they could be summoned in case of emergency by electric bells. The brigade was equipped with a steam-engine, the boiler being always laid ready so that steam could be got up in 15 minutes once the engine was

63 Many wives prepared dinner for their men ready to be collected from the houses by carriers' carts and brought to the Market Place. From there they were picked up as the men emerged from the gates and were eaten in the dining hall or in one of the pubs with a pint

on the move. The police at the factory, a detachment of N division Metropolitan Police, comprised three sergeants and nine constables, all of whom lived on the factory site and could be summoned in the same way as the firemen. Meantime the shafting in the large room had deteriorated so badly that it had to be entirely relined.

The dining hall, in 1906, was made more comfortable by the installation of steam heating. An annual grant of £12 had been paid to the Tottenham Hospital on which the factory had chosen to rely in case of serious accident. On several occasions however it had been found impossible to get attention there and, since the Enfield Cottage Hospital had on these occasions proved obliging, it was recommended that the grant should be transferred there. The RSAF Provident Society now had 2,386 members including 342 former employees. It had been set up in 1885 and thoroughly reorganized in 1903. The benefit was now £14 on the death of a member and £7 on the death of a member's wife. A levy would be automatically imposed should the fund fall below £1,000.

Outings were arranged by each department in the summer. On one of these, described by Sylvia Collicott, 85 men from the smithery travelled in four brakes to Bishops Stortford where lunch was taken and the men played cricket, watched a game of polo and rounded off the day with a musical evening organised by Mr

Martin the assistant foreman.

The Factory Institute continued to provide a lending library, a newsroom, lectures and entertainment. A grant of £40 had been made to the Middlesex County Education Committee for the provision of classrooms at the Ponders End Institute for the instruction of trade lads from the factory, they attended one afternoon a week, in the firm's time. The course fee of £1 was to be refunded to the boys on completion of the course. Workmen instructing the lads were to be given 2s a week additional pay. Trade lads had been recruited for many years, for example in 1886 Col Arbuthnot had taken six, chosen by competition from among 48 candidates.[18]

The Ordnance Factory at Sparkbrook in Birmingham closed in 1907. Costs at Sparkbrook had always been higher than those at Enfield and, unlike Enfield, the factory there could not be expanded. The closure necessitated provision for the instruction of armourers at Enfield. The system for the recruitment and training of armourers' lads had been set up in November 1903. The boys were selected by the officer commanding Army Ordnance Corps, they came to Enfield at the age of sixteen. Their course extended over 2½ to 3 years. Each day began at 8am with outdoor gymnastics, running and musketry. They were trained in the factory and in the evening they attended school. They received pay and allowances amounting to 18s a week and were provided with board and lodging in selected houses for which 13s was deducted from their pay, this left 5s for pocket money. Their uniforms were of course provided.

Until this time the works had maintained its own sewage disposal system. It was now decided that this should be abandoned and that the factory sewers should be connected to the Council sewerage system.

Much of the repair work, which had formerly been carried out at Sparkbrook, had now to be transferred to Enfield. The land, factory and machinery at Sparkbrook was sold to the Birmingham Small Arms Company which proposed to use the works for the manufacture of the Short Magazine Lee Enfield. The Government had induced the BSA to buy by guaranteeing them a proportion of the orders placed with the trade. Two private firms were capable of making the rifle, the BSA and the London Small Arms Company, but there was no competition, and their prices were invariably the same. It had always been the intention of the War Office to encourage the growth of the interchangeable system in the private sector and it was acknowledged at BSA that the experience gained at Enfield had been freely made available to the firm, models and gauges had been supplied without restrictions. The skills learned from Enfield were (as Putnam and Weinbren have demonstrated in their *Short History of the RSAF*), developed by private arms firms when they subsequently diversified production into other fields in engineering, as in sewing machines, bicycles and motor cars.[19]

The introduction, in the early years of this century, of high-speed steel brought about changes in machinery, which resulted in a speeding up of work known as

64 Maxim machine-guns at the RSAF 16th May 1905

65 Armourers lads putting on a display at the factory in 1908

'speed and feed'. This was to be organized by Col F. Nathan who was appointed superintendent of both the RSAF and the Powder Mills in February 1909. Nathan was ordered to give special attention 'to the economies aimed at in the attached proposals'. He was given a free hand to rationalize production and cut the number of employees. He met with stubborn opposition from the Small Arms Union and the Amalgamated Society of Engineers. The War Office refused any longer to guarantee the maintenance of 2,000 jobs at Enfield and the numbers now fell to 1,900. All men over sixty had been discharged, and all those under twenty-one. A great demonstration was organized, supported by five to six thousand people with banners and bands, to protest against the 'sweating methods being adopted and the heavy reductions in wages at the Royal Small Arms Factory'. There was much concern throughout the town that any serious reduction of the wages paid at the factory would hit both shopkeepers and landlords. A protest meeting held at the Bycullah Athenaeum, in a very upper-middle-class area, was addressed by Henry Ferryman Bowles, the former Conservative Member of Parliament for Enfield, J. R. Pretyman Newman, the current Conservative Member and his Liberal rival in the January 1910 election, James Branch.[20]

The National Insurance Act 1911 at last provided RSAF workers with a measure of protection against unemployment and sickness. Unemployment benefit was paid at 7s a week for 15 weeks in any one year. Those earning up to £160 a year were entitled to 10s a week sick pay and free medical treatment from their

66 *The Enfield Highway Co-operative Society's new central stores opened in Ordnance Road opposite Titchfield Road in 1907. The society thus retained a link with the men at the factory and their families*

ROYAL SMALL ARMS FACTORY

ENFIELD LOCK.

A

MEETING

WILL BE HELD ON

Thursday, August 12th,

AT 12.20 P.M.,

To Discuss the following Resolutions:—

1. **GOVERNMENT EMPLOYEES.** — That this Meeting calls upon the Government to give definite expression to the model employers' resolution, passed in the House of Commons in 1893, by fixing the minimum rate of pay (30/- per week) for all Government employees and thereby ensuring a living wage.

2. That all piecework prices be settled by a joint Conference of men and officials of each department concerned.

3. That all employees of the Government be placed upon the same basis as all Government Officials as regards Pensions and Holiday Pay, by giving them at least one week's holiday in the Summer with pay, also all closing down days, namely, Stocktaking, Bank Holidays, &c.

4. That no Pension Scheme can be considered just or adequate that does not include all Government Workers having served seven years or more in any Government Factory at any time preceding the adoption of the scheme now before the Government.

5. That no contract for ordnance work be given outside any Government Factory until all such Government Factories are brought up to their proper normal strength of employees in the time of peace, and that the cruel practice of discharging young men at 21 years of age and old men at 60 years of age, be at once discontinued.

6. That the Government be called upon to enforce immediately the Fair Wages Clause, passed in the House of Commons, 1908, upon all Contractors at present doing Government work.

7. That a copy of the foregoing (by the vote of this Meeting) be sent to Mr. HALDANE, M.P. (*Secretary for War*), JAMES BRANCH, M.P. and all the Labour Members in the House of Commons, asking their support.

C. J. HURD (T.U. Throughout), New Road, Lower Edmonton.

67 A protest meeting to be held 12 August 1909, presumably in the Market Place outside the factory

68 The years between the Boer War and the Great War were years of uncertainty for the workers. The War Office strove to reduce labour costs and there was a constant threat of redundancies. Things grew desperate and, at the beginning of April 1910, a great demonstration was organized supported not only by the unions but by local councils, the Board of Guardians and local shopkeepers. It was attended by more than five thousand people

panel doctor. They would contribute 4d a week. This gave rise to a dispute with the unions, for the men already paid 6d a week to cover sick pay and treatment by Dr Priest, the factory doctor.

Some indication of a premonition at the War Office can be seen in the report (August 1911) that the machine-gun shop was again working at high pressure following years of slackness.

Prospects in May 1913 hung in the balance, and seemed to depend upon the adoption or rejection of a new service rifle which fired rimless ammunition. It was about three inches longer than the current design and had a peep-hole sight instead of the 'v' shaped. Retooling for its production proceeded slowly, both because of the limitation on the number who could be taken on in the tool room, and because no firm orders had been received. But in August 1913 preparation for its production at Enfield was stopped although thousands had already been spent. Certain weaknesses had developed, the new rifle was described as unserviceable. There was the promise of a new machine-gun (the Lewis) more deadly than the Maxim and so light that one man could carry it. The rationalization had left the work-force disgruntled; it was a 'speed and feed' system, each work process being watched and timed by the management. The staff in the sword and

69 Ordnance factory workers from all over London supported the protest

bayonet department had been cut by half and those remaining were mostly boys. The unions contested the savings made and criticized the quality of the work produced. Union effort was now concentrated on securing a 30s minimum wage for labourers, while the ASE sought an advance of ½d an hour for its members at RSAF, to bring them in line with other engineers in Enfield; at Rees Roturbo, Ediswan, Metallic Flexible Tubing and the Nobel Explosive Company. J. E. Mullins,

70 A great demonstration against the new 'sweating methods' of 'time and speed' was discussed at a meeting in the parish hall, 650 workers were crammed inside while 150 waited outside. J. H. Matthews urged that each shop should adopt a banner

72 The smithy had 86 fires, 20 steam hammers, 8 screw drop hammers, 15 forging machines, and 3 sets of rolls for making swords and bayonets

71 (opposite) Workers assembled in the Market Place. On the left is the lock house, on your right the factory gates and the school, beyond is Government Row

73 A solemn service held to mourn the death of King Edward VII, 6th May 1910, this illustration has been taken from a postcard, posted on the 19th of that month. Much of the pool (the mill head) was filled in, presumably during the First World War, to make space for more workshops. The mill tail had been lost between 1865 and 1885 and a new grindery and proof-house had been erected on the site

the union representative, pointed out that there had been no pay increase at the RSAF for 12 years previous to 1913, but the War Office remained adamant in its refusal. Indeed many of the older hands found themselves discharged or transferred from the workshops to what they called the 'labour gang' doing unskilled work at reduced pay. A new process for drilling rifle barrels was introduced in 1914 and many barrel setters, who had been earning £2 a week, were put on other work and found their wages reduced to 26s. The factory was in the unhappy position, as late as March 1914, of being equipped to do work and having the prospect of little to do. The dismissals were bitterly resented. One Saturday, after the factory had closed, men of the smithery assembled to honour two of their comrades who had lost their jobs. Unrest continued, rumours swept the factory that large-scale dismissals would follow the installation of labour saving machinery. Why, it was asked, should the factory be confined to the manufacture of arms when motor transport could be made, or aircraft? The Birmingham Small Arms, under private enterprise, and Armstrong Whitworth, had both gone into motor production. Obviously war was not foreseen, except perhaps in Germany from where it was reported that military experts had tested a new device for mounting quick-firing guns on aeroplanes far in advance of anything hitherto known.[21]

Colonel Tisdall resigned as chief inspector of small-arms at the factory early in the year 1914, to be replaced by Major Brown. Government estimates for the factory for the year ending 31 March 1915, announced in May 1914, foretold no worsening nor improvement in the prospects; the allocation for wages remaining the same at £200,000; for materials less money was allocated, £75,000, against £80,000 in the previous year. The Government was certainly not preparing for a war that year.[22]

Notes to Chapter 5

1 C 5116. 219, 357, 7
2 *Gazette* 6S, 28 Oct 1889
3 *Gazette* 27 D 1889
4 PP 1888 lxvii, 1890 xliii, *Gazette* 6 Mr, 16 Ja 1891, Barrass corres.
5 SUPP 6. 651
6 *Gazette* 8 Mr, 25 Mr, 1 Ap 1892, PP 1893/4 Lii 691
7 SUPP 6. 651
8 *Gazette* 15 S 1893
9 *Gazette* 2 D 1892, 3 F 1893, 15 Jl 1892, 7 Ap 1893, SUPP5. 126
10 *Gazette* 10 Mr, 24 Mr 1893, 7 Oct 1892, 20 Jl 1894
11 *Gazette* 20 Mr 1896
12 *Gazette* 17 Mr, 19 My, 10 N 1899, 12 Ja 1900
13 *Gazette* 26 Ja, 19 Ja, 26 Mr, 6 Jl 1900
14 *Gazette* 25 My, 15 Je 1900
15 *Gazette* 18 Ap, 4 S, 11 S, 25 S 1903, AEU Enfield Lock Branch
16 Cd 3626, *Gazette* 4 Mr, 1 Ap, 3 Je 1904
17 *Gazette* 3 F 1905, PP 1907 X, *Gazette* 26 My, 18 My 1906
18 SUPP 5. 168, PP 1887 xiv 364, PP 1907 x 421, SUPP 5. 689, SUPP 6. 658
19 SUPP 6. 655, Cd 3626 (ie PP 1907 X)
20 Putnam and Weinbren *op cit*, *Gazette* 18 Ap, 17 Je 1910, SUPP 6. 658, *Gazette* 17 Je 1910, 25 Au 1911
21 *Gazette* 23 My, 26 D 1913, 13 F, 6 Mr, 20 Mr, 24 Ap 1914
22 *Gazette* 6 F, 15 My 1914

Chapter 6
The Great War

The war scare which agitated the country in July 1914 was intensified in Enfield by the presence of troops at Enfield Lock and Waltham Abbey to guard the government factories. The Royal Garrison Artillery with pom-poms (machine-guns), and men of the King's Royal Rifles, had taken up positions at the Gunpowder Works. A contingent of the Somerset Regiment arrived at Enfield Lock by the 11.15pm train and was quartered for the night in the school-room at the factory. The following day a strong body of police was drafted into Enfield Lock and lodgings were sought for them, indicating that they had come to stay. Night-shifts were being worked in some departments at the RSAF by August. Men worked a twelve-hour day, or a twelve-hour night-shift, and soon a thirteen-day fortnight became compulsory.[1]

The RSAF was better prepared to increase output than might have been expected, for the factory had already been geared for the change to mass production. The new projected rifle had to be abandoned and Enfield continued with the Short Magazine Lee Enfield. Over two million of them were made there in the four years of war. Although the Lee Enfield rifle was produced in many parts of the world including the USA, Enfield was proud to be its place of birth and its home.[2]

From the early months of the war the number of men of military age working in the factory aroused criticism. Captain Pretyman Newman, the local Member of Parliament, speaking in Enfield in October 1914, described his constituency as 'one of the worst places in the country' when it came to providing volunteers for the Army. It had been hoped to raise locally another battalion of the 7th Middlesex Territorials, but a great many men, who either had, or hoped to get, jobs at the Small Arms Factory or in other war industries, were hanging back. Army pay was certainly no incentive to induce a man to join up. The wife of a private soldier with three children could expect to receive only £1 a week, unless her husband made over a portion of his own pay. Since many factory workers were working overtime, and doing night-work, pay at the RSAF was much more generous. A great deal of betting went on and bookmakers recruited runners from among the employees. Herbert Stevens, a labourer of Durants Road, employed at the RSAF, was arrested outside the factory 'with gold, silver and slips' and was ordered to pay £10, with the alternative of a month in prison. The four public houses at

74 There were floods at the end of December 1914. 'By Wednesday morning', wrote the Enfield Herald, *'the water had risen to three feet and the Small Arms Factory stood out in the centre of a large lake. Two thousand men on night duty were completely cut off ....' Many had to wade almost waist deep across the flooded area, only to find their houses under water.*

Enfield Lock were now allowed to remain open until midnight and until 12 mid-day on Sundays. The clubs in the area were also exempted from the general prohibition on drinking after 11pm in order to accommodate the men at the RSAF.[3]

Drunkenness at work was nevertheless strongly discouraged. One worker was suspended for a month for this offence, and since the Metropolitan Munitions Tribunal refused to discharge him, to allow him to apply for work elsewhere, the punishment was equivalent to a fine of £20. He complained that he had never before been cautioned and that the usual suspension was for one week only. But excessive drinking must have been causing problems for the management, because the Liquor Control Board, in December 1915, suddenly decided upon the drastic restriction of opening hours in the four public houses around the factory. On a Saturday night just before Christmas, the night-shift, coming out for their meal break at 11pm, found all the pubs closed. The men were angry, Colonel Halse the deputy superintendent was called to the scene and went some way to placate the workers by granting them a two-hour break so that those living nearby would have time to go home, or to get something to eat in the factory dining hall. Many went home and failed to come back that night. On the Sunday, in consequence of the licensed houses not being opened until 1pm, the men again had to be granted an additional hour. The new regulation thus resulted in a loss of eight to nine thousand hours work. On the Sunday evening a branch meeting of the Workers Union was convened and a delegation, led by McGrath, was sent to the Liquor Control Board. As a result the four houses were allowed to remain open from 11pm to 12 midnight, but not for the sale of intoxicants. McGrath argued that his union had no wish to see the pubs abused by the workmen to the detriment of production, but there should be facilities for men leaving off work at 6 o'clock in the morning, and those starting at 6.30am, to get a little refreshment. In the evening the houses should be allowed to open at 5.30 not 6.30 to cater for those starting or coming off at 6 o'clock. On Sundays, although the men came out at midday, the public houses opened only from 1pm until 3pm McGrath suggested what common-sense should have proposed, that the opening times should be 12noon until 2pm. The temperance lobby still seemed to work most effectively to the greater inconvenience of the common man. Soon afterwards, on 2 February 1916, the four houses were placed under government control. Two of them were to be converted into canteens.[4]

The management continued to recruit but refused to take on men aged between 19 and 35. Close on 5,000 were employed by the end of 1914. There was a grave shortage of housing accommodation in Enfield and many had to travel long distances to get to work. Their difficulties were increased by the inefficiencies of the tram service along the Hertford Road. Frustration was inevitably aroused by the sight of three to six trams, in convoy, mostly empty, proceeding blindly on their way to Waltham Cross while long queues built up along the road behind them. Public transport however was greatly improved in April 1915, both by the provision of a better service on the trams and by the reopening of the Cheshunt Loop for passenger trains. The improvement was short-lived for in May a strike broke out among LCC employees at the Wood Green tram depot and it quickly spread to Tramway Avenue. The men declared that they would carry munition workers only, and sought to gain the sympathy of the workers at Enfield Lock by a march to the Ordnance Factory where they explained their grievances at a meeting in the Market Place during the dinner break. It was addressed by Henry Barrass, now a councillor in Edmonton. A further improvement in public transport came in June 1916 when a new station (with a wooden platform) was opened on the north side of Carterhatch Lane to serve passengers using the auto-train service on the Cheshunt Loop.[5]

King George V and Lord Kitchener visited the Ordnance Factory and the Gunpowder Factory in May 1915. This led to a surge of enthusiasm among middle-class ladies in the area, eager to involve themselves with the welfare of the war workers. They set up a committee to provide refreshments, non-alcoholic of course, for the night-shift in the early hours. Tents were erected for the lady caterers within the factory compound. Working conditions were improving, men got each

75 Middle-class women sought a role for themselves in the war effort. Most of them had never before even had to clean their own homes or cook meals for their families, now they were to be found making tea and providing food for the night-shift in the early hours

alternate Sunday off instead of one a month. A 3d return workmen's train was provided by the Great Eastern Railway from Liverpool Street to Enfield Lock, the difficulties of railway management however required that it should arrive there at 6.15am although the shift began work at 7.[6]

Jobs at the RSAF were much sought after and each morning crowds of men stood waiting in the hope of work, often as many as two hundred were turned away. Nevertheless there was still a shortage of skilled labour and those with skills, but not engaged in the war industries, were requested to register at the Public Offices in Gentleman's Row as War Munitions Volunteers. The minimum rate for unskilled workers at the RSAF in August 1915 was 6d an hour, which amounted to 27s for a 48-hour week, but men were working as many as 82 hours a week and, beyond their normal hours of work, they were paid time and a quarter for the first two hours overtime, and then time and a half; double time was paid for Sunday work.[7]

The Times, on 22 September 1915, carried a letter concerning a war bonus for workmen at the RSAF, it cited a claim for damages by a worker whose wages were £7 10s a week. Labourers, the writer complained, who before the war were earning 25s, were now averaging £3 a week, skilled mechanics much more, and not a penny direct tax had been levied. 'Lads leaving school at 14', he alleged, 'are now getting 18s a week, more than double their normal pay'. The unions denied that £7 10s was a typical wage, merely an isolated example earned on piece-work, moreover many labourers, it was claimed, averaged only 29s a week. Another report concerned a single man earning £4 a week at the RSAF. He was released to induce him to join the Army, but he went to Erith and found employment at the same wage at Vickers machine-gun factory. The Gunpowder Workers Union at Waltham Abbey, in October 1914, amalgamated with the Workers Union at the RSAF.[8]

Housing and lodgings remained difficult. Scarcely a bed, let alone a room, could be found in Enfield and the Government was soon expected to engage a further thousand men at the RSAF. To remedy the accommodation shortage the YMCA, in August 1915, began building 60 huts on a six-acre meadow east of the Lock railway station, part of Welch's farm. The huts would be divided into cubicles, each containing a bed and an easy chair; bathrooms, dining facilities and a large recreation room were to be provided. Accommodation for eighty workmen was ready by September, they were to pay 17s 6d a week to include laundry. Mrs Winston Churchill promised to officiate at the opening in October. There was a serious shortage of dining space near the factory, the old dining hall, the Sheridan Theatre, had a fully equipped kitchen where the catering was let out to contract. Another dining room was opened in July 1915 but the number of men employed at the factory was increasing so rapidly that there was chaos at meal times, both day and night. Men were reduced to eating their food standing in corridors, sitting on stairs or herded together in small rooms in the public

76 To house the great influx of workers, sixty huts like these were erected in a meadow near the railway station in August 1915. They were divided into cubicles each containing a bed and an easy chair

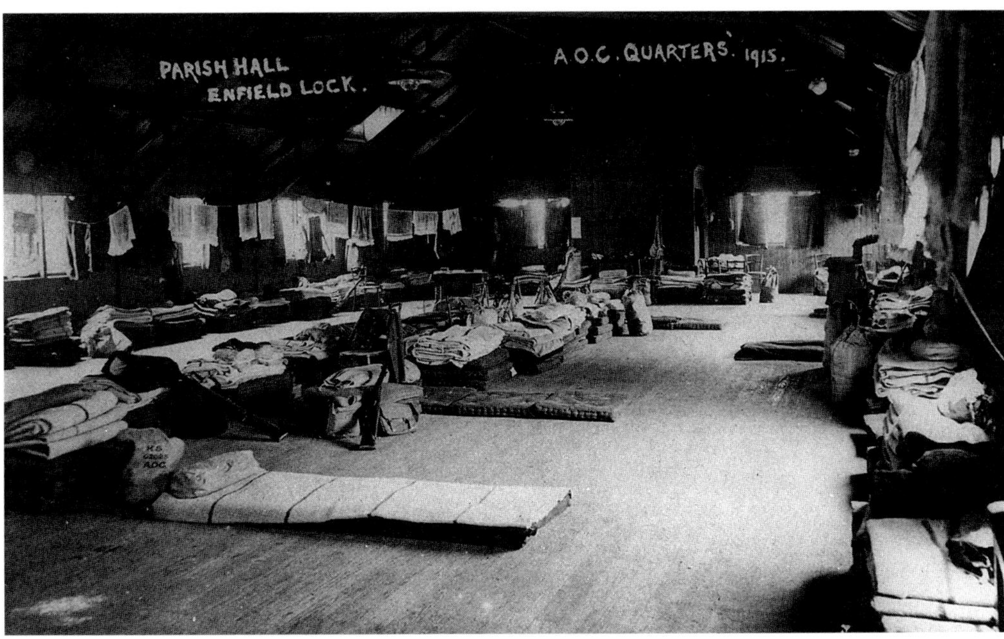

77 Army Ordnance Corps personnel billeted in the parish hall Ordnance Road 1915

houses. Those that were not fortunate enough to find even such shelter, some three-quarters of them, were forced to take their breaks in the open. The Liquor Control Board, in consultation with the management at the factory, therefore decided to acquire (nationalize) the four public houses and to enlarge the Greyhound and the Royal Small Arms transforming them into canteens. Work on the Greyhound was completed by 27 November 1916 when it was formally opened by E. S. Montagu, Minister of Munitions. The house now had 350 sittings, was open from 5.45 in the morning and, throughout the day and night, one thousand meals, breakfasts, dinners and suppers, were prepared. It could now serve many times the number of customers as formerly. Temperance advocates were delighted to find, although their reasoning may have been perhaps dubious, that while the number of meals served increased sevenfold the sale of intoxicants had *merely* doubled. Twelve months later the Royal Small Arms Tavern was completed on an even larger scale; it was opened by the Duke of Connaught. The Swan and Pike and the Ordnance Arms were given minor improvements. The changes were welcomed by the men and an address was presented by them to the Chairman of the Control Board. A committee of the workers was formed which sat at intervals to advise and consult with the general manager of the Liquor Control Board.[9]

Work was in progress in October 1915 on a siding to bring wagons from the main line at Brimsdown into the factory, the line was carried on an earth bank across the marsh between the Turkey brook and the Old River Lea, three bridges were required. The siding, costing £36,630, was to prove of inestimable value. By 1917 as many as seventy trucks a day were unloading in the works, not only goods for the RSAF, but bulky goods like coal, acetone, acid and cotton waste brought in to be transferred to Royal Gunpowder Factory (RGF) barges and towed on by a motor launch especially purchased for this purpose.[10]

Many of the younger unskilled workers at the RSAF registered in November 1915 as willing to join the forces under Lord Derby's scheme at any time they should be called upon. It was the Liberal government's final effort to avoid conscription. Hands were constantly being taken on although preference was given to discharged soldiers and men over military age. A number of skilled men, wearing khaki, had been recalled from the front. The factory buildings were extended northward following the establishment of the Ministry of Munitions in July 1915. A hundred more cubicles were proposed for the RSAF and a hundred more at Ponders End Shell Works. Fifty female operatives were taken on to work machines in the large room in April 1916 and a canteen for them, with a recreation room attached, was erected in Government Row. There a piano was provided and the lady superintendent organized dances and gymnastic classes. Wounded soldiers were invited to the entertainments, the women having taken responsibility for the men in a ward at the Waltham Abbey VAD Hospital. It was planned to employ 1,500 women at the RSAF and 1,000 at the RGF. Between five and six

78 A large canteen was built for the women during the First World War and run by the management. It had a recreation room attached and a piano was provided. Dancing classes and gymnastics were arranged and wounded soldiers invited to the entertainments

79 The YMCA canteen during the First World War

80 Women workers wore blue overalls and a cap to protect their hair from the machinery. Between five and six hundred were employed at the factory by August 1915

hundred were employed at Enfield Lock by August 1915, mostly on machine work. They wore blue overalls and a cap to protect their hair from the machinery. The Workers Union urged that women's pay should equal that of men. The increased numbers at the factory enabled the Ministry, in May 1916, to discontinue Sunday working, for which it had to pay double time.[11]

In January 1916 the Military Service Act imposed conscription for the first time in this country and local tribunals were set up before which a conscientious objector was allowed to state his case, as well as any man who claimed exemption on the grounds that his work was of national importance. The Cheshunt tribunal protested that two thousand eligible single men remained employed at the RSAF. David Weston, a Liberal, who was chairman of the Enfield Town tribunal, lent his weight to the protest. The Edmonton tribunal even went on strike over the matter, claiming that single men had replaced married men at the RSAF. In consequence a deputation from the tribunals, accompanied by Sir Charles Ellis of the Ministry of Munitions, visited Enfield Lock. Col Fisher, superintendent at the RSAF, pointed out in answer to the criticism that all engagements and discharges had to be approved by himself, or by his deputy, he assured the delegation that favouritism or influence in the retention of jobs was not possible. With this the tribunals were only partly satisfied. Henry Barrass, a member of the Edmonton tribunal and an official of the Workers Union, continued to stir up trouble and the undercurrent of ill feeling persisted. Rumours were spread that money was paid weekly by some young men in order to maintain their jobs and badges. The tribunals maintained their campaign and in December David Weston was able to express his pleasure that efforts to get single young men out of the factory had at last met with success. Orders had been given to de-badge all unskilled and semi-skilled workers under 41.[12]

Between August and November 1916 the number of women employed at the RSAF increased from 575 to 983, in the same period those employed at the RGF increased from 387 to 1,145. Since de-badging, 395 men had been released from the RSAF and 516 from the RGF. Yet there still remained 1,220 single men of military age at Enfield Lock of whom 535 were classified as indispensable and 250 as unfit, including a number of discharged soldiers. The remaining 435 were subjected to growing pressure to force them to leave.[13]

Wartime building extensions, either temporary or permanent, from the beginning of the war up to February 1917, had cost £114,725. These included a new stocking shop, extensions to the machine-gun and hardening shops and to the barrel mill, a new automatic screw shop, a dining hall, kitchen and other facilities for the women, a new saw mills, a new browning shop and assembly shop, and new office accommodation. Machinery had been purchased on a large scale at a cost of £156,620, there were new machines for the manufacture of Vickers locks, machines to equip the barrel plant and for the manufacture of bolts and caps, machines to equip an extension to the bayonet shop and machinery to increase the output of rifles and machine-guns, and to equip the new stocking shop. Electric power had been laid on from Brimsdown Power Station and a supply of gas from the Tottenham and District Gas Co. A new footbridge had been built over the Lee Navigation replacing the one constructed in 1883. Further structural alterations, which promised ‘ a greatly enhanced appearance to the main entrance’, were made that February and some old buildings were demolished and further offices built.[14]

In July 1914, before the outbreak of hostilities, a thousand rifles a week were being made at the RSAF; by July 1917 weekly production had reached 10,500. In addition to these rifles, which were produced entirely at Enfield, 2,500 a week were assembled from components manufactured elsewhere. Work on rifle repair fluctuated according to the number returned from the front, varying from 7,000 to 9,000 a week. All the gunstocks, 30,000 a week, were produced at Enfield Lock. A large mill had been built to deal with the timber, but the shortage of walnut caused anxiety. Continental sources had been cut off at the beginning of the war and purchases had to be made in America, despite a prejudice which existed against American walnut. Various parcels of walnut had been discovered in this country, notably at H. Lebus in Ferry Lane Tottenham, but by July 1917 the RSAF had in hand only sufficient for eight weeks production. Thirty-eight weeks further supply was ordered from America. Some 30,000 stocks a week were

81 During the Great War many men cycled to work. A bridge for cyclists had been constructed over the level-crossing in Ordnance Road in 1909

82 The day's work done and a long procession of workers heads homeward along Ordnance Road

required. Early in the war there had ben a great demand for cavalry swords, this fell off gradually, though not surprisingly, and had by this time ceased.

Altogether 7,040 men, 1,095 boys and 1,448 women were employed in June 1917. The women and boys had been taken on to replace men released to the colours. Nevertheless in view of the highly specialized nature of much of the work wholesale dilution was not possible. Out of 7,820 males over sixteen, 1,836 were single, of these, and excluding the scheduled men and discharged soldiers, only 51 were under 31 years old and classified 'A'. By this time 1,470 men had been released to the Army of whom 280 were married and 1,190 were single.[15]

Labour relations had remained stable throughout the war up to 1917, despite the fact that a pay award in 1913 had left the engineers employed at the RSAF with a grievance. The ASE at that time had secured an advance of 1½d an hour for its members in Enfield, but because the Ordnance workers in the London area were tied by a five-year agreement, members working at the RSAF had to be excluded, so that their pay would remain in line with engineers in London Ordnance factories. Since the beginning of the war two bonuses had been awarded; 3s in November 1914 and 4s in March 1915, in line with other Ordnance factories. In April 1916 grievances were again causing discontent. That month a mass meeting was held at which it was claimed that, from being the best paid factory in Enfield, the RSAF had become the lowest paid employing ASE members. There were further meetings of the engineers in September and October and on Sunday 5 November a meeting was held at the Liberal Hall at Enfield Wash. Strike resolutions were suppressed by the chair but seven hundred highly skilled men sought an advance in pay. The quality of their work, it was claimed, had to be higher than average, for they were mostly employed making gauges for the factory and master gauges for contracting firms. The men demanded a 10s increase and time and a half for night-work. They were granted only a further 3s bonus. Unrest and agitation among engineers in government factories continued to spread through the Lea Valley until, in May 1917, six hundred RSAF men downed tools. Four hundred engineers in other Enfield factories came out in sympathy. The management held that the engineers were adequately paid. The minimum rate for a mechanic at the beginning of 1917 was 51s 6d for a 48 hour week. Mechanics on the day-shift however worked 57 hours, including 9 hours overtime, and the average wage was £3 4s 6d. Men on the night-shift did 66 hours and their average wage was £4 17s 4½d. The majority of the men worked alternate day-shift and night-shift. Thus even for the lowest paid mechanic, claimed Col Fisher, the increase in pay over pre-war was 50 per cent but for the majority it was 100 per cent. The ASE pointed out that purchasing power was now only 45 to 50 per cent of pre-war.

Hostility to the strike by the unskilled and semi-skilled workers was voiced by J. McGrath, secretary of the Workers Union in the factory. Meetings against the strike were held, one at the corner of St Stephen's Road and one in Ordnance Road, from there demonstrators made their way to the YMCA huts east of the station, where they demanded that all the strikers should be turned out. The strike was short-lived, it was over before the end of May.[16]

Floods at Enfield Lock in January 1918 had turned the area into a vast lake. The factory could only be reached by wading, work had to be suspended at dinner time and carts turned up to carry the workers. The Greyhound was under two and a half feet of water.[17] Easter that year was a stay-at-home holiday, with the heavy fighting in France the arms and munition workers got no respite. The machine-gun department was very busy. Ill feeling about jobs at the RSAF remained rampant. The Discharged Servicemen's Association held a protest meeting outside the factory, speakers accused fit, single young men of sheltering there and preventing 'those who had done their bit' from getting jobs. The meeting also found fault with the employment of young girls. A large proportion of the appeals before the Enfield tribunals were still made by workers at the RSAF and the RGF.

Further extensions to the works were made in June. A considerable area of the Market Place, up to the gable end of the school, was taken into the factory; the changes considerably reduced the space which had been available for stalls. At the top end of Government Row increased accommodation was added to the recently erected offices.[18]

There were rumours of peace in October. A notice was put up in the factory announcing that in view of diminished output, and in order to save coal, the factory would close from 1pm Saturdays, until 7am Mondays. It was on Monday 11 November that a maroon fired from Enfield Town Police Station foretold the imminent end of the war. At 11 o'clock came the official notice by gunfire that the war was over. The RSAF closed down and everyone came out to celebrate.

Before the war the RSAF had been turning out 1,000 new and 1,000 repaired

83 Floods in Ordnance Road in January 1918 — looking west (towards Hertford Road) near the junction with Salisbury Road

rifles a week. Within three months production had increased to 3,000 new weapons and 8,000 repairs. Production rose the following year to 6,000, twelve months later to 9,200 and by 1917 to 14,300. During 1918, 6,700 Vickers machine-guns were repaired.[19] The present entrance to the factory was completed in November.

Notes to Chapter 6

1 *Gazette* 31 Jl 1914
2 E. G. B. Reynolds, *The Lee-Enfield Rifle* p180
3 *Tottenham and Edmonton Herald*, 18 N, 25 S 1914
4 *Gazette* 15 Oct, 10, 17, 31 D 1915
5 *Gazette* 18 D 1914, 9 Ap, 21 My 1915, 26 My 1916
6 *Gazette* 7, 28 My 1915
7 *Gazette* 11 Je, 2 Jl, 25 Je, 6 Au 1915
8 *Gazette* 25 S 1915, 30 Je 1916
9 *Gazette* 23, 30 Jl 1915, PP 1917/18 XV 684, PP 1918 XI 117
10 *Gazette* 15 Oct 1915, MUN 4. 4894, MUN 5.151.1122.1/3
11 *Gazette* 5, 19 N 1915, 21, 28 Ap 1916, MUN 5.151.1122.1/3, *Gazette* 12 My 1916

12 *Gazette* 25 Au, 29 S, 13 Oct, 22 D 1916
13 *Gazette* 22 D 1916
14 *Gazette* 12 F 1917, MUN 4.4894, MUN 5.151.1122.1/3
15 *ibid*
16 SUPP 6. 661, *Gazette* 18 My 1917
17 *Gazette* 18 Ja 1918
18 *Gazette* 5 Ap, 28 Je 1918
19 *Gazette* 18 Oct, 15 N 1918, 9 My 1919

CHAPTER 7

Between the Wars

On 22 November 1918 the *Enfield Gazette* warned its readers that many thousands employed in local munitions factories would, before long, find their occupations gone. At the RSAF notice had been issued to announce the end of overtime and the resumption of a 48-hour week, Monday to Friday. Some, including many over-age men, had already been given two weeks notice, now all employees were granted three days holiday with pay. The YMCA was much concerned as to what should be done with a thousand huts which would inevitably become redundant. By 31 January demobilization of labour had led to the dispersal of thousands of workers, it was rumoured that by the end of March numbers at the RSAF would be reduced from 12,000 down to a pre-war 2,000. Floods in February added to the gloom. The waters extended westward almost to the YMCA hostel meadow just east of the railway line. Many on their way to work turned back. The more enterprising bethought themselves of the elevated branch line from Brimsdown into the factory premises, their enterprise however was in vain for, the back river continuing to rise, all work had to be stopped at 9am and the works remained closed on the following day.[1]

Although the outlook for the RSAF workers looked bleak the closure of the factory was unlikely, but it was ordered to concentrate on repair work only. The Deputy Minister of Munitions had made the situation clear in reply to a question in Parliament in April: 'on account of the large stocks, both of small-arms and small-arms ammunition,' he said, 'the War Office demand is so small that it is not possible to place any fresh contracts with firms in the Birmingham district... both from the point of view of economy and for the maintenance of a permanent staff at Enfield,' he continued, 'it is essential that the repair of rifles should be concentrated at that factory'. Some 120 tons of rifle components and 170 tons of Lewis parts, with all the necessary gauges and tools, were in consequence removed from Birmingham to Enfield to facilitate any repair work which might arise. The men who had been engaged on rifle and machine-gun manufacture at BSA had by this time been discharged. The Army Council could not foresee any demand for new small-arms and therefore refused to support a recommendation by the Minister that a nucleus of staff for the *manufacture* of rifles should be maintained at Enfield, indeed the supply of rifles from store was to remain adequate up to 1939. It was thought moreover that, since the Lewis gun was unlikely to be retained as a standard weapon, stocks of these made available from return by the troops would be adequate to meet any possible contingency. Colonel Halse urged that Enfield must store all the tools and gauges in case at any time it should be necessary to resume output on a large scale, especially since the BSA was now no longer prepared to provide storage.[2]

Easter 1919 was indeed a contrast with the previous Easter. Numbers were reduced from a wartime 12,000 to 2,700 and the factory stood closed for stocktaking, but employees now received a minimum wage during such enforced idleness. A mile away to the south the Ponders End Shell Works closed and all hands were discharged. During the war the works had employed 6,000 on day and night shifts. To the north the Royal Gunpowder Factory was threatened with closure; 5,000 men, women and girls had worked there throughout the war. The pessimism around Enfield Lock was increased by a report that a machine-gun factory was being built at Burton-upon-Trent.[3]

Through the autumn of 1919 the machinery which had been used for mass production lay idle. In the large room only 15 machines were being worked out of 1,245, in the sight department only 11 out of 574, in the machine-gun shop 3 out of 323, in the barrel shop 4 out of 552, in the bayonet department 39 out of 259, in the screw department 20 out of 300. In total, out of 3,467 machines, only 161 were being operated. The factory generators were closed down in June.[4]

Despite the times, apprenticeships in the Royal Ordnance factories remained in great demand. Entrance was by examination, there were 163 candidates for 43 places in July 1920 and 10 were secured by students at the Ponders End Trade School. The requirements of the War Office for small-arms remained low. In September piece-workers were stood off for half a day a week, leaving work at midday Friday until Monday morning. The following January the men, not surprisingly, turned down a proposal that they should work short-time in order to absorb more of the unemployed, despite this a scheme was imposed by which they lost one week's work in six. Apprentices, it was ordered (August 1922) should be discharged on completion at the age of 21 and re-engaged where desirable for temporary employment.[5]

Fear inspired by the Bolshevik Revolution caused the Government to overreact to any threat of industrial action. It rushed into emergency measures when the miners came out on strike in April 1921, erecting barbed wire entanglements around the Royal Small Arms Factory and posting troops to guard the perimeter. There was of course no violence but the strike did result in a shortage of coal and the workers were put on short-time.[6]

That January railway wagon repair work had been introduced at the RSAF. Some railway wagons were built at Woolwich at a cost of £15,000 each, but although they were offered for sale at £10,000, they remained unsold. The repair work at Enfield was at first done in the open, but subsequently a large building was adapted and railway lines were laid. Orders however did not materialize and

84 The factory church closed January 1922, one of the economies imposed by the 'Geddes Axe'. The building was demolished 1928

the scheme had been dropped by August, a few men were transferred to other work but 230, mostly taken on since January, were discharged.[7]

It was announced early in November 1921 that the RSAF church must close. The War Office grant of £200 a year was withdrawn in the economies at that time imposed upon the War Office. The little church had stood within the factory compound since 1846, approached from Government Row by a wooden bridge over the back river. The church was closed in January 1922.[8]

Facilities for making tea in the workshops, introduced during the Great War, were withdrawn that same month and a number of foremen, about sixty years of age, received notice. The 'Geddes Axe' was but beginning its work, but its full effect was to become apparent when the government estimates appeared in April. Mr Carnegie, not long since appointed superintendent, was transferred to Woolwich in February 1922, G. H.Roberts was placed in charge both of the RSAF and the RGF. Nevertheless £5,000 a year was being spent on the repair of machines in the vacated shops. A request was made to increase the rate of repair from two to four machines a week.[9]

Unrest existed among the unskilled workers and the Workers Union called a meeting during a dinner hour. The labourers had a grievance over wages, they were paid £2 4s 6d a week at a time when even a council labourer was paid £2 13s 6d. They were given a 3s increase but this still left them, according to W. W. Henderson, the new Labour Member for Enfield, below the poverty line. Even skilled men, he urged, were not able to save against the day when they would be discharged at between 60 and 65, and then they would receive only a small gratuity. Their grievance over superannuation had lingered on for 54 years. A mass meeting was held at the end of April 1924 in the Assembly Hall in Ordnance Road under the auspices of the Enfield Lock and Waltham Abbey branches of the Workers Union. The men demanded an additional 10s a week. Royal Ordnance labourers, with the 3s increase, now received 47 19s, but the pay of council road sweepers by this time had increased to 51s. Skilled workers were awarded 6s a week from February 1924 but the labourers received no further rise. Discontent continued, J. McGrath, the Workers Union secretary, pointed out that, at RSAF, semi-skilled men were paid only £2 9s while outside workers of a comparable status received £3 9s. Management however, forced to make stringent economies, had become less susceptible to pressure. A memorandum had been issued in March 1922 giving notice that the trade unions must not interfere with the exercise of management functions, the memorandum required the unions to have regard for the *Provisions for Avoiding Disputes,* 17 April 1914.[10]

Small-arms repair work was exhausted by 1922 and it was ordered in April that the personnel at the RSAF was to be reduced by 600 to 1,350. The pre-war strength had been 1,800. At the RGF numbers were to be halved to a mere 250, compared with between 800 to 900 before the war. The Waltham Abbey Traders Association asked that the cuts be deferred, and organized a meeting, calling together representatives of the unions, the local councils and the Board of Guardians. The *Gazette* spoke of a threatened calamity in eastern Enfield and complained that the increase in local unemployment, made worse by the continued discharges at the RSAF and the RGF, was beginning to affect trade. Members of the Geddes Committee in May 1922 recommended the scrapping of the RGF, pointing out that the country had a ten-year supply of explosives. The consternation aroused locally by this proposal gave rise to a question in Parliament from the Enfield member T. Fermor Hesketh. The Waltham Abbey factory remained open, but the great munitions factory at Gretna, which had cost £9 million to build during the war, was sold off piecemeal in January 1924; at this time there was scarcely any work going on at Waltham Abbey. The sacking of men continued and, by the end of May 1922, 530 men had left the RSAF while the RGF had been reduced to a nucleus of 230; only those most experienced in the manufacture of cordite were retained, these therefore included very few ex-servicemen. Most of the two hundred factories built by the Ministry of Munitions during the First World War had by this time been closed. Private munitions factories continued to be wound up, Nobel's cartridge factory at Waltham Abbey closed in February 1925 with a loss of 1,500 jobs. The work was transferred to Birmingham.[11]

Nevertheless the management at Enfield Lock sought to plan for the ensuing ten years. Proposals, it was said, to make the RSAF 'an ideal factory', would be welcomed. It was pointed out however that the time was not opportune for big schemes, for little money was likely to become available. But there remained many matters which could be considered for action, such as electricity and gas supply, water mains, the standardization of machine-tools, jigs and gauges, and mechanical coal handling. It was judged that the factory ought to retain its own gas plant, for its production costs, over three years, was estimated to be 3s 6d per 1,000 cubic feet, as against a price of 4s 9d from the Tottenham and District Gas Co. The factory gas plant was extended in 1924.

A design department was established in 1922 with a library of trade catalogues. It led to the production at the RSAF and at other factories of new weapons like the 0.45 and the 0.38 inch revolvers and the Mk VI SMLE (rifle No 4). Manufacture could now be done from fully dimensioned drawings, obviating the expense of producing hardened metal gauges.[12]

The number employed at the RSAF was reduced again in September that year to 1,214, a loss of a further 85 jobs. Numbers had fallen to 1,184 by September 1926 and a further reduction to 1,100 was expected by 31 March 1927. Most of the discharged men, it was pointed out in the *Gazette*, would have to wait two or three years to get their old age pensions (10s a week), in the meantime they would have to find their full contribution and they would be forced to fall back on a meagre unemployment benefit for subsistence.

The factory was divided into sections at this time, each under a manager. Bennett Powell was shop manager east, in charge of the assembly, browning and carpenters' shops, the hardening shop, polishing, repair and sight shops, and the stocking branch. A.Heath, the shop manager west, was in charge of the barrel shop, the large room, the bayonet and pistol shops, the screw shop together with the

85 The Telpher track used to convey coal to the boiler house. The coal heap can be seen and a coal barge on the canal

smithery and the foundry. A. Amies controlled the machine-gun room and the toolroom. Rate-fixing estimates, planning and instruction were the responsibility of J. Hodson. W. Crocker managed the examination branch.

Many changes were being considered including the future of the Enfield Standard Inch. The principal clerk offered to investigate 'its ancient history'. The Enfield Inch, which had been devised in 1860 to ensure greater precision, was to be replaced from October 1924 by the British Standard Inch. Two of these old Enfield standards were to be retained for workshop use and two for reference, one to be held in the factory and one by the Army Inspection Department. An agreement was reached with the LNER empowering the North Metropolitan Electricity Supply Co to take over the siding from Brimsdown which had been built to bring supplies into the RSAF during the war. Items of machinery considered for purchase in the year 1924/5 included a condensate separator, a magnetic lifter, and a weighbridge. It was proposed in December 1924 that, providing a supply of castings could be secured, the foundry would no longer be needed. The smithy should be retained but in a modified and concentrated form with a heat-treatment and welding shop added. This work was carried out at a cost of £791. The smithy stacks were in such a dangerous condition that they had to be demolished. The wharf cranes, no longer used, were disposed of. Much to be regretted was the demolition of the large room porch.[13]

A proposal was made to the War Office in November 1924 that, four machines a week should be refurbished, as against two a week in the current year. In August 1925 it was agreed further to increase the number of machines being put through for repair. The main purpose of this work seems to have been the provision of employment for skilled men in the toolroom, that they might be retained in readiness to install plant for the Mk V1 rifle (Rifle no4). Five machines a week were being repaired by November. The *Daily Express*, in March 1926, published an attack on the inefficiency of our 'bow and arrow arsenals', declaring that Enfield was twenty-five years behind the times and still using machinery from the time of the Indian Mutiny.[14]

From time to time skilled men were transferred to the labour gang, to work as latrine attendants, window cleaners and road sweepers. It was ordered in August 1925 that such men must be returned to production at the end of three weeks so that the disadvantages of work in the labour gang would be spread over the largest possible number of men. A Hospital Savings Association was formed at the RSAF in July 1925. Members paid 3d a week to insure themselves against the expense of hospital treatment.[15]

The Webley Pistol was being produced at the factory, its cost being progressively reduced from £3 8s to £2 18s 4¾d. A threatened shortage of work was relieved by an order for 2,500 pistols in April 1926, but the lack of work at the factory was making it difficult to maintain discipline. Early morning inspection by management had found men idling in the polishing and browning shops, and men starting late in other shops. The principal clerk was asked to press the War Office to settle on the new pattern rifle and new pattern revolver, so that employment could be found for gauge makers who otherwise would have to be discharged. Once such skilled men were lost their replacement might prove especially difficult because of the local housing shortage.

The Government again took somewhat excessive security precautions at the time of the General Strike in May 1926. A detachment of soldiers was posted to protect the factory. Some 800 RSAF men joined the strike but no one was penalised when the strike was over. The *Gazette* claimed afterwards, with what justification I don't know, that many union members at the RSAF 'have thought well to tear up their trade union cards'.[16]

The erection of machines in the new machine shop (shop no 1), the former large room, which reopened in June 1927, proved costly. Plant and machinery in the year 1928 was purchased to the value of £25,000. The aim was not to achieve immediate increased output, but to make some provision against the contingency of a future war; the expenditure was said to be the first instalment of between £100,000 and £150,000 for the purchase of machinery and equipment. The weakness of this scheme was that new machinery could not be placed in its permanent planned position unless the building reconstruction programme was accelerated, which would mean that money in the draft estimates would have to be divided between machinery and buildings. Believing that this would cause too many difficulties Roberts decided to devote the entire £25,000 to machinery. Nevertheless, in 1928, he sent the Department of Finance his plan for the future layout of the factory which might be worked to over the ensuing years. He had originally devised this plan in 1922, aiming then to forecast development up to 1932. The plan showed a decrease of 103 in the number of separate shops; of these 33, which were wooden buildings and so had constituted a fire risk, had already been demolished. The plan did not show any large projected developments such as a new machine-gun department which was likely to be required in preparation for any new war.[17]

Much experimental work was carried out on the development of the No 4 Rifle, the former Mk V1. Many of the earliest of these weapons, unsatisfactorily produced elsewhere, presumably from accurate drawings supplied by the RSAF, had been brought to Enfield for inspection and rectification. Tests were also carried out on alternative designs for components. Early in 1926 six prototypes had been manufactured for trials and shooting tests at Enfield. The new design had passed technical trials by 1930 and manufacture had begun of the first 500 which were expected by May 1931. But the year was one of severe cuts in defence expenditure, public opinion remained strongly opposed to rearmament. The No 4 Rifle was not put into mass production until after the outbreak of World War II. The work was then done at two new Royal Ordnance factories; Maltby and Fazakerley and at (BSA) Shirley. A number of Enfield men, skilled in the manufacture of rifles, were to form the nucleus of the work-force at Maltby and Fazakerley. Enfield by that time was engaged almost entirely on the production of the Bren light machine-gun and the Enfield .38 pistol revolver.

The 1930s began badly with the abolition of piece-work, in consequence 500 men (160 skilled and 340 unskilled) faced a loss of earnings amounting to between 13s and 20s a week, the Transport and General Workers Union sent a deputation to discuss the heavy loss of wages, but there were two million unemployed. It was at this time that shop stewards at the RSAF and the RGF sought management recognition for shop stewards' committees to negotiate pay and conditions. Nevertheless men continued to lose money by transfer from piece-work to time-work, and in September 1931 the labourers were forced to accept a wage cut from 49s a week to 46s. AEU members at both the RSAF and the RGF called a meeting to oppose the cuts which also threatened them, but by November seven to eight hundred men faced further loss of earnings. Men were now discharged on reaching the age of 61, and 150 semi-skilled men lost their jobs in February 1932.[18]

The long-drawn-out campaign for pension rights, lost in 1870, still continued and in February 1933 J. McGrath led a deputation to the War Office. There followed a further deputation in March when Duff Cooper rejected the union's demand for 10s a week pay increase. Exclusive of a 12s a week war bonus, mechanics were earning 46s 6d, machinists 36s 6d and labourers 34s, for a 47 hour week. A few men received higher rates because of the nature of their work, but certainly RSAF workers were not better paid than men in other Enfield factories.[19]

J. McGrath JP retired in July 1933 after 40 years service at the RSAF, although he remained secretary of the local branch of the TGWU until April 1938. He was a native of Birmingham, of Irish parentage and had come to Enfield in his early twenties, in 1890, when, as he said, 'there was neither house nor light between Ordnance Road and Waltham Cross'. In 1903 he became secretary of the Factory Union of Small Arms Employees, for at that time each factory had a separate union, but in 1915 it amalgamated into the Workers Union and, in 1928, into the Transport and General Workers Union. For a long time the union could get no recognition from the War Office. It was Lord Haldane who, in 1909, spoke in the House of Commons in favour of recognition, wages had then been 7s to 10s for a 54 hour week.[20]

It was announced in August 1935 that the Lewis gun was to be replaced by a light automatic Czechoslovak designed machine-gun. The War Office had reached agreement to use the patent and would manufacture the weapon at Enfield. The new machine-gun, it was claimed, would have greater portability, lower cost, and would also be 'suitable for the cavalry'. The Enfield-Brno collaboration produced the Bren gun, a .303 light machine-gun. Output at Enfield was set to increase rapidly. One serious result of the decline of small-arms production in the UK through the inter-war years was the failure to design new weapons; the Bren, Browning and Besa machine-guns, together with the 20mm. Hispano and Oerliken, were all of foreign design. The only weapon of small-arms calibre originated at Enfield during the rearmament years 1936-9 was the Boys anti-tank rifle. After 1939 the RSAF staff designed both the Sten light machine-gun and the Polsten, a 20mm. weapon for the Army.[21]

German troops marched into the Rhineland demilitarized zone on 7 March 1936. Hitler now felt free to rearm. That month the Defence Requirement Sub-Committee recommended a substantial expansion of the Territorial Army and advised the creation of 'shadow factories' which were to be encouraged by small orders for arms, to enable them to finance expansion and to purchase new machinery. A few months later the Rome-Berlin Axis was launched and recruiting began at the RSAF. Ossie Edwards began work there at this time at the age of fourteen, in the stocking shop where the butts were made for the famous Lee Enfield rifle, with other wooden parts. He worked a 49-hour week, including Saturday mornings. His pay, as a member of a piece-work team, averaged about 16s a week. The first task allotted to him was to fit a nose-cap to the front handguard using two small screws, for this the piece-work price was 9d a hundred; his target, which he rarely achieved, was 300 a day.

Another of his jobs, a more fulfilling task, was to make the wooden handgrips for the old-fashioned sword-type bayonets. Given a piece of American walnut, he carried through all the operations; copy-turning, drilling and sanding, to produce the finished handgrip. All the wooden parts produced in the shop, following inspection, were taken to the dipping shed and immersed in linseed oil for 24 hours. He still recalls with pleasure the pungent smell that pervaded the place. Between the age of 14 and 16 he was engaged in a variety of repetitive jobs, the most tedious being that known as 'bullets'. This involved the removal of the pip left on the steel bullet after it came off the automatic lathe by pushing it through a hole in a face-plate onto a revolving cutter, then twisting and pulling it out. He received 5d a thousand and innumerable cuts and grazes to his hands.

On his sixteenth birthday a boy could apply to become a trade apprentice. Ossie Edwards became an apprentice millwright. Although in this job he would earn less than a production worker, he learned his trade from men who had, in the main, used basic hand tools, the tools of a craftsman; hammer, chisel and file. He remembers seeing an old millwright forging 2 inch by ½ inch steel to make a circular bracket, perfectly round, and showing not a single hammer mark. The old tradition of craftsmanship in the factory was based upon men like these.

Although push-buttons for the motors in machine-shop 1 had been installed in 1929 most of the machine-tools at Enfield remained belt-driven. With the provision from the Northmet of A/C current from Brimsdown however, and the purchase of three 750kW diesel generators over the years 1939, 1940 and 1941, the replacement of belt-drivers was speeded up, although it was not finally completed until after the Second World War.[22]

Leslie Hore-Belisha, Minister at the War Office, visited the factory in January 1939 to see the manufacture of the Bren gun which was in full production on a wartime scale. Enfield Lock was the main source of supply of Bren guns in this country. In two years up to January 1939 the number employed at the RSAF had risen from 900 to 3,400. Advertisements for girls appeared in the local press:

> Female tracers, preferably between 16-18, required at RSAF ... previous experience in an engineering drawing office desirable; salary from 20s 6d to 34s a week according to age and experience.

The rate of rearmament continued to increase. The Army was to be expanded to 32 divisions, six regular and 26 territorial.

The provision of up-to-date plant for the large-scale production of the Bren was a most important development. Deliveries of the weapon had begun in September 1937 and full production was reached during the year 1938. To keep the

86 (opposite) Most of the machines remained belt-driven in 1936 but, with the provision of AC current from the Northmet and the purchase of three 750kW diesel generators around the beginning of the war, more individual electric motors could be introduced which gave greater flexibility in machine speeds and reduced both danger and noise within the factory. The illustrations show the conversion of the stocking shop in 1955

W131
W97

Stocking Shop

87 Leslie Hore-Belisha, Secretary for War 1937-40, inspecting Bren guns at the RSAF, behind him T. Denness the superintendent

work-force happy the War Office at last arranged for a private insurance company to provide pensions for the workers. The Ministry of Supply was established in April 1939 and conscription was introduced.[23]

Notes to Chapter 7

1 *Gazette* 22 N 1918, 17 Ja, 21 F 1919
2 MUN 4 6699, Hornby, W. *Factories and plant*, HMSO 1958
3 *Gazette* 4 Ap, 28 Mr 1919, MUN 4.5334
4 *Gazette* 28 N 1919
5 *Gazette* 23 Jl, 17 S 1920, 21 Ja 1921
6 *Gazette* 8, 15 Ap, 13 My 1921
7 *Gazette* 29 Jl 1921
8 *Gazette* 4 N 1921, 6, 13 Ja 1922
9 *Gazette* 10 F 1922, SUPP 5. 947
10 *Gazette* 24 Mr 1922, 6 Ap 1923, 28 Mr, 6 Je, 5 Jl 1924
11 *Gazette* 14 Ap, 12 My, 21 Ap, 9 Je 1922, 20 F 1925
12 SUPP 5.947, 1 Au, 9 N 1922, 12 N 1924, Putnam and Weinbren *op cit* p.95
13 *Gazette* 1 S 1922, SUPP 5.947, 8 Mr, 3 Oct 1922, 14 My, 17 D, 5 Au 1924, 8 Ap 1925
14 SUPP 5.947, 12 N 1924, 12 Au, 11 N 1925, *Gazette* 19 Mr 1926

15 SUPP 5.947, 12 Au 1925, *Gazette* 31 Jl 1925
16 SUPP 5.947, 11 N 1925, 22 Ap 1926, 12 N 1924, 7, 14, 21 My 1926
17 SUPP 5.947, 26 Je 1927, 30 Mr, 18 Je 1928
18 *Gazette* 21, 28 Mr 20 Je, 22 Au 1930, 20 Mr, 11 S, 20 N 1931, 5, 12 F 1932, Putnam and Weinbren p101
19 *Gazette* 20 F, 31 Mr, 7 Ap 1933
20 *Gazette* 28 Jl 1933
21 *Gazette* 2 Au 1935, Hornby, W. *Factories and plant*, HMSO 1958
22 'British Machine Tools' in *Engineering* Ap-Je 1947, SUPP 5. 947
23 *Gazette* 27 Ja 1939, *Enfield Herald* 28 Oct 1938, *Gazette* 16 F 1940, 27 Ja 1939

CHAPTER 8

The War Years & After

Most armies began the Second World War with the same rifle systems, even the same rifles, as they had used in the 1914-18 war. There had been no demand for rifles in the UK since 1918 and it was now found that, for the first time for more than a hundred years, the country was left without a substantial reserve capacity for rifle production. In consequence it was agreed that the No 4 Rifle had to be the British standard service rifle, mainly because it was the most suitable weapon for mass production. Manufacture however could not begin in this country, even in 1939, for it had to await the construction of

88 King George VI firing the Bren on a visit to the factory in 1940

89 Machine shop 2, making small components for the Bren, probably wartime, 'Music while you work' through the tannoys

90 Brens being test-fired on an indoor range at the RSAF

new factories. Meantime the No 4 was produced in Canada and the United States.

At the outbreak of the war, machine shop No1 contained a thousand machine-tools making components for the Bren. Vickers machine-guns and components for the Hispano-Suiza were being made on 500 machines in machine shop No 2. In machine shop No 3, 300 machines and 150 men produced the .38 revolver, machine shop No 4 made barrel and other components for a variety of weapons. The smithy contained drop hammers and steam hammers and three furnaces. The repair of rifles was carried on. The toolroom and the small-arms inspectorate had a role in setting-up the manufacture of various small-arms in other factories.[1] According to Ossie Edwards working practices at Enfield changed little at the outbreak of the war. It was the fall of France that brought about a new feeling of urgency and the tempo of work increased at once. To replace the weapons abandoned on the beaches across the channel required a national effort, a system of seven consecutive twelve-hour shifts was introduced at the RSAF.

The industrial area from Ponders End north to the Royal Gunpowder Factory, especially the power station and the RSAF, offered tempting targets for the Luftwaffe, indeed a German wartime aerial photograph exists which clearly shows the factory. The general stores and a new building housing the gauge room, as well as the drawing office built on a platform spanning the 'middle river' were

91 Iron tunnel range used for weapon testing, note the extraction fan above the firing point

largely demolished by a land mine which fell at two o'clock in the morning on Sunday 21 November 1940, although the nearby water-tower and gas holder were left standing. One workman was killed in a subsequent raid when a small bomb came through the smithy roof and exploded near the men while they were having their meal break.

The factory had its own air-raid precautions (ARP) and, after 1938, staff were trained in a building just north of the water-tower. In the expectation of immediate and heavy bombing, shelters, constructed of steel panels covered with earth, had been sited on open ground on the east side of the factory. Each man was allocated a particular shelter, forty personnel to each, and a marked route was designated from every workshop.

During the autumn of 1940 air-raids increased and warnings sounded almost every night forcing night-shifts to shut down their machines and take shelter. Several hours might then elapse before the all-clear. To continue night-work under such circumstances was not possible, yet some means had to be found to maintain production. Shelters of solid brick with concrete roofs were therefore

92 Damage caused by the land-mine which hit the factory 21 November 1940. This photograph was taken from the water-tower.

put up adjoining each machine-shop and on the top of the water-tower a look-out was erected. A factory warning system using large light globes and bells was introduced, but was only put into operation when immediate danger threatened. Work could now be abandoned at the last moment and resumed when the immediate danger had passed. It was fortuitous that on the night of the land-mine the warning system had come into operation and the night-shift had been suspended, for the explosion left scarcely a window or a roof light in the factory, casualties otherwise would inevitably have been high.

The RSAF Home Guard battalion (56th Essex) was well armed with Brens, Stens, rifles and mortars. The ranges at the factory were made available to the unit, mostly young men, who were exempt from military service due to the class of work they were employed on. According to Ron Evans (he was a corporal dispatch rider) they were the envy of other units, less well provided.

Shortly after the outbreak of the war the Pattern Room was moved to Broxbourne, the wartime headquarters of the Chief Inspector of Small-Arms (CISA). From there patterns (no longer formally sealed) were issued as reference guides for mass production and were vital in maintaining output, particularly of

93 The same incident taken from the gas holder

the Sten gun. The Pattern Room co-ordinated the work of hundreds of small engineering companies keeping them up to date with design changes and improvements by sending out sample components and drawings. The Pattern Room returned to Enfield in 1948 where it remained until the closure of the RSAF in 1988. It is now housed at Royal Ordnance Nottingham.

From the beginning of the war the numbers working at Enfield continued to expand to reach 6,315. Women were employed, often at skilled work as examiners and viewers. Shift working was used from the beginning of the war. Many cycled to work, there was no bus service in Ordnance Road until late in the war. Wages rose rapidly and differentials, between men and women, and skilled and unskilled, narrowed. Peggy Smith who worked in the office at the factory during the war earned £10 a month plus £3 4s 9d overtime and thought she was rich. She recalled that one worker on the payroll 'knocked up £15' in a week on piecework. 'He was stopped from working so fast', she said, 'because he was setting a rate that others couldn't do'. According to Ron Everton and Bill Belcher people

94 Several such pill-boxes remained on the factory site after closure. About seven feet tall, they were erected to enable WD police to take shelter while patrolling during air raids. The apertures were filled with five or six inch thick glass

would spend the firm's time making cigarette cases and lighters, brooches or spoons, one man mended spectacles. Nevertheless output at its highest reached 10,000 Brens, 1,000 Stens and 1,000 pistols a week.[2]

There were dances in the canteen and concerts by ENSA. The canteen was popular. 'The food was good', said Peggy Smith. Before they had official tea breaks workers reheated their own tea in the bucket of hot water under the bench (used for rinsing the finished product). Later they were allowed to brew up in the workshop, but afterwards tea had to be brought from the canteen.

The expansion at Enfield Lock was part of a comprehensive scheme of rearmament. Plans for twenty-three new Royal Ordnance factories had been approved

95 (opposite top) A look-out was constructed on top of the water-tower

96 (opposite bottom) Sadly the Pattern Room has been taken from Enfield Lock and is now housed in Nottingham. It might well have formed the nucleus of a museum telling the glorious history of the Royal Small Arms Factory

97 Women working in the canteen kitchen during the Second World War. The kitchen was well-equipped with steam, gas and electric ovens, hotplates and fish fryers. There was a buffet adjoining the canteen where tea, coffee, sandwiches, cakes and biscuits were served. Trolleys were taken around the factory for morning and afternoon tea breaks

by September 1939 and four were already in operation, eighteen others began work in 1940. The processes of manufacture of small- arms had to be simplified as far as possible into elementary repetitive tasks. These processes relied heavily on the skill and versatility of the Enfield toolmakers in making the jigs and gauges on which the production of the workshops depended. Equally vital was the careful planning in sequence and the close control of the machinery operations.

Concerned with small-arms were the Royal Ordnance factories at Fazakerley, Maltby and Poole where the first superintendents were men trained and experienced at Enfield. From here too, were transferred many highly skilled staff who were to form the nucleus of the management. The success of these factories through the war is a tribute to the men from Enfield Lock. One 'Lockie', who went north early in the war to impart his skills, was Thomas Slingo, he became foreman in a Royal Ordnance Factory at Warrington in Lancashire where he worked until 1946. Born at Enfield Lock, he had joined the Kings Royal Rifles at the age of 17, was

98 Barrel making during the Second World War, the setter (in overalls) looks on

wounded and invalided out in 1917, and later became an inspector at the Royal Arsenal.

Fazakerley, an entirely new factory on the outskirts of Liverpool, began the production of small-arms in June 1941, Maltby in Yorkshire began work about a year earlier. The prototypes of all the weapons they produced came from Enfield whence also came the skilled instructors to teach the new semi-skilled labour force. At the peak of its operation Fazakerley employed 12,000, mostly on the manufacture of the Hispano gun for aircraft, 70 per cent of the labour force were women. Throughout the war 750,000 of the number 4 Rifle, which had been designed to be made in large quantities using unskilled and semi-skilled labour, were produced there. Also manufactured there was the Sten, designed at Enfield by Shepherd and Turpin; the initial letters of their surnames ST with the EN for Enfield, provided the name Sten. Its design was so simple that components could be made in small engineering workshops, even garages, from detailed measured drawings supplied from Enfield. A shortage of tools caused difficulties but many engineers, with the help of toolmakers from Enfield, contrived to make their

99 Barrel room families workers' outing 1949

own. These two factories, with the parent company at Enfield Lock, produced during the Second World War, five million small-arms including Stens, Brens, Hispano, pistols and rifles. Major-General John Hay Beith described the expansion of the Ordnance factories from three to forty-four as 'one of the greatest, though naturally least advertised, feats of organization achieved in the Second World War'. At the peak of their activities in 1943 about 350,000 were employed, most of whom were women. The first need had been for instructors, they came from Woolwich, Enfield and Waltham travelling over the country like missionaries.[3]

No armament orders were placed by the Government with the Ordnance factories after the end of the European war in 1945. The factories had to survive on

alternative work from private industry, but this was both uncertain and unsuitable. The Enfield factory produced such things as carpenters' tools and household plumbing equipment for the many building workers being released early from the forces to facilitate post-war housing reconstruction. The alternative work programme was not a success. Enfield had a tradition of high quality and high costs, perhaps essential in the manufacture of complex arms but less appropriate in making tools for builders or household goods. There could be no attempt to sell on an open market. 100,000 New World cookers were made at Radway Green ROF, but all for one customer, said G. Dyer. Jack Gibbons, looking back on this period recalled a beautiful carpenter's mallet, of exceptional quality. The care put into the production would only have been justified for something far more elaborate.[4]

G. R. Strauss, the new Minister of Supply, stated in October 1947, that it was government policy to maintain the labour force at the Royal Ordnance factories at about 40,000. He agreed that any declarations of redundancy at Enfield and Fazakerley, about which workers' representatives had expressed particular anxiety, would be withheld until January. These representatives had put forward a case for using the full capacity of the factories, after having met the needs of the forces, for civilian production. They were of the opinion that the factories were capable of making capital goods such as mining machinery, pointing out that these were both badly needed in this country, and could make an important contribution to exports. They contended that, without any sales organization, the Ministry of Supply had received, since the end of the war, firm orders for £20.5m worth of civilian work. Their estimate of work completed amounted to £7m.

Government policy in regard to the Royal Ordnance factories was again outlined by G. R. Strauss in February the following year. 'For their war potential', he stated, ' as for their economic usefulness to the nation, we must keep the factories as active concerns with a healthy complement of workers. The permanent establishment of some proportion of the workers is under consideration. My purpose therefore', he went on,' is to seek for these factories every kind of alternative work that can be properly undertaken. In this connection we intend to take a stand at the British Industries Fair to attract new orders'.

The number employed at the factories at this time was 39,600 and it was likely to continue at much the same level, although this did not mean that there would be no fluctuations.

Despite the alternative production a Commons select committee in April 1949 complained that two-thirds of the capacity of the Ordnance factories remained unused, either for service or for civilian work. The Ministry of Supply, reaffirming that its policy was to maintain the nucleus of manpower for making weapons in case of emergency, pointed out that, in the engineering factories such as Enfield, workers could only be kept active, either by increasing the demand from the forces, or by creating space for civilian production, but this would diminish the capacity of the factories to make arms in such an emergency.[5]

Weapons designers at Enfield developed the first 'bull-pup' rifle, the EM2, in 1948. It was a revolutionary system subsequently used in the SA80. It was recognized at the time to be a potential world-beater but in 1952 Winston Churchill halted its development in order to provide work for the post-war recovery of the European economies. He had ruled that small-arms for the British Army should be purchased from a Belgian company, Fabrique Nationale, whose 7.62mm rifle, the FN SLR, was thus adopted by NATO.

Many of the Ordnance factories in August 1949 were still engaged in a desultory fashion in civilian production. That morale was not high can be seen by the workers' attitude to sickness benefit. Since this had been made up to full pay, for up to thirteen weeks in one year, there had been so marked an increase in the number of cases of reported sickness that J. H. Jones of the Ministry of Supply was forced to visit each factory, seeking the reason and explaining the nation's difficulties at meetings of the employees.[6]

Five years of retrenchment, redundancies and alternative production continued until June 1950 when the situation was radically changed by the war which broke out in the Far East. North Korean forces, armed with Russian weapons, invaded South Korea and there was strong pressure for rearmament. The Cabinet increased defence expenditure from seven to 10 per cent of national income and it was further augmented in February 1951 to 14 per cent. At this time National Service was extended to two years. The impact of rearmament engendered feverish activity in the Ordnance factories, efforts had to be made to convert them back to full-scale production with all possible speed. The transformation was required at once. Every order was subject to a short-term delivery. Fortunately, four years earlier, although the time had then seemed unpropitious, it had been decided to examine the condition of machine tools, plant and equipment at the Enfield factory and a modest start had been made on a scheme for replacement, though only in so far as the stringent financial limits then in force had allowed. The RSAF was therefore in a better condition than might have been expected. Recruitment of labour now became a major problem, particularly at Enfield, for heavy recruitment was required. In May 1951 the Ministry of Supply announced that fifty per cent more labour would be needed by Britain's 21 Royal Ordnance factories to carry out work on the rearmament programme. The current issue of *ROF News*, the official house organ of the factories, stated that between 15-20,000 workers would have to be found as production increased. As many women as men would be required. Women were being sought for light engineering, but there was a particular shortage of skilled men, especially machine-tool operators, fitters, electricians and carpenters. Trade unions were approached concerning the reintroduction of dilution, as during the war, but this was strongly resisted on the shop floor, as were all attempts made by management to achieve greater productivity per man-hour. Management was disturbed by the amount of time lost settling minor labour disputes, especially those concerning piece-work prices.[7]

A decision was taken in 1952 to transfer all Hispano production from Poole to Enfield. It provided yet further incentive to the re-establishment at the RSAF of a capacity for quantity production of both small-arms and the light cannon gun, enabling Enfield Lock better to compete with other Royal Ordnance factories and with the trade. The requirements of rearmament now enabled the factory to acquire new modern machine-tools. But problems for the management recurred whenever new techniques were introduced for it became necessary to persuade the workers, or their shop stewards, that neither loss of earnings nor redundancies would ensue. The speeding up of work was linked in the minds of the workers with the fear of unemployment. Management faced an incessant struggle to retain their co-operation, for it could only be done by steadily increasing the average hourly earnings on piece-work. Yet every advance in piece-work prices created unrest among the hourly paid workers and it was found difficult to get young men or former apprentices to undertake day work. All new prices introduced were looked upon with suspicion by the unions and disputed as a matter of principle. Stubborn resistance was offered to any proposal to transfer men from one machine to another, or from hand work to machine operation. Thus the introduction of new machinery or new processes came always to be met by last ditch resistance from the unions and it became virtually impossible to secure union agreement to the introduction of further semi-skilled labour.[8] These were views attributed to 'senior personnel' by H. W. Hobbs in the ROF annual report for the year 1953. Many former employees however dispute these opinions saying that they give a distorted view of industrial relations. 'During the Sixties', at a time when he was factory secretary, Bert Hunt, 'found that the majority of industrial and non-industrial representatives were men who were keenly interested in the reputation of the factory'.

The Korean War ended with the armistice of July 1953 and the division of Korea along the 38th parallel. Despite the end of the war the number employed at Enfield Lock grew from 1,483 (3 March 1954) to 1,553 a year later as the

100 View of m/c II in 1956 before machinery was installed for production of the FN rifle

Cold War intensified. Production work proceeded at both Enfield and Fazakerley and orders were received for the FN rifle and the Patchett carbine. Great efforts were being made to put the Patchett carbine into production. Work study was being considered.[9]

After the period of 'excitement and excesses' of the Korean rearmament the factory at length returned to the more sober atmosphere of normal peacetime demand. The first prototype of the FN rifle was issued to the services in 1955 and the re-laying of the workshops for its production was begun. Relations on the shop floor continued unsatisfactory, workers' representatives were determined to maintain earnings at levels at least equal to those earned during the rearmament boom, although men were now producing less because many were working on new and unfamiliar jobs. There was, according to the management, a concerted trade union attack on shop discipline and the exercise of managerial functions. Despite the fact that an increased percentage of production went for export, the Board of Ordnance now faced difficulties in maintaining full employment at both

101 View of m/c II after machinery was installed

Enfield and Fazakerley. At 31 March 1956, 1,497 hands were employed at Enfield Lock, the year's production was valued at over £1m and £108,000 had been spent on plant and machinery including 65 machine tools.[10] Orders had been issued for 50,000 FN rifles in February 1956; 12,000 at Enfield, 18,000 at Fazakeley, the remainder at the BSA. The RSAF was required to produce 22 FN rifles for testing by the end of May 1957 and, following successful trials, the total order was increased in July to 250,000; from the ROFs 100,000 and 150,000 from BSA. The weapon would be .303 inch calibre, it would therefore have to be converted for use by NATO which had adopted the 7.62mm calibre round. Production planning for the FN rifle (adopted by NATO in 1954) was transferred from Fazakerley to Enfield in October 1957. The weapon was described in the *Scotsman* at the time as 'rather an ageing debutante in the Army's family of weapons'. It had certainly been around some years. Prototypes had first been produced in Belgium before 1950 and it was subsequently adopted by the Government for the British forces.[11]

The successful testing of a British hydrogen bomb at this time however foretold a diminishing role for the RSAF. Aubrey Jones of the Ministry of Supply spoke in Parliament of the reshaping of the defence forces then taking place. He was unable to say what capacity in the way of ROFs would be needed, though clearly some contraction was inevitable. He promised that the factories would retain a vital role in our defence, being regarded as the preferred source for the sorts of munitions each was equipped to produce, he warned however that in special circumstances the Government might have to look to alternative sources of supply. He also promised to bring increasingly to the factories work on the newer types of weapons, but he also warned that it was unlikely to lead to large-scale production, or to the employment of large numbers. He promised to keep the factories in an efficient state and whenever military demands fell off, and when the capacity of the Ordnance factories could usefully supplement private capacity, they would be allowed to undertake suitable civil work. The changes in defence policy were outlined in the White Paper on Defence in April 1957. Aubrey Jones, speaking to representatives of the ROF workers in May, again insisted that there must be a process of contraction and it must take place as quickly as possible. He was not yet in a position to name the factories that would have to close. He had met trade union officials and representatives of the factories at a special meeting to discuss the future in the light of the White Paper on Defence. It remained Government policy that the factories existed to meet the current needs of the forces and to provide, in any emergency, a reserve of arms-making capacity. Until recently however the emergency in mind had been a major war lasting several years, but now, if a major war should break out between the great powers it would be both devastating and short and there could be no question of massive wartime production increases. Nevertheless limited wars, like those in Korea and

Indo-China (Vietnam) might still have to be provided for and for these the factories must be the preferred source of production.

Where however ROFs, kept in being, could not be fully employed making arms he again emphasized that it was reasonable for them to undertake civil work to maintain the appropriate labour nucleus. Britain's defence would henceforth depend largely upon the hydrogen bomb, delivered it was hoped by a fleet of Vulcan bombers or the Blue Streak missile. Conscription was to be terminated by 1960 and by 1962 the Army would be reduced from 700,000 men to 165,000 professional soldiers.

The atmosphere of urgency now diminished at the RSAF. There was a proposal in February 1958 to reduce the FN order to 200,000 and to spread the work over a longer period. It looked as though there would be virtually no work either at Enfield or Fazakerley after the year 1960/61, for by that time the work on the conversion to 7.62mm. and the work on the Patchett carbine would be completed.[12]

Nevertheless the numbers employed at the RSAF continued to grow until the end of the Fifties. There were 1,770 employed on 31 March 1958 and a year later 1,828, but during that twelve months five Royal Ordnance factories had been closed including Poole, From there, work on the manufacture of spares for the Aden 30mm cannon gun, and a complete set of tooling was passed to Enfield which then became responsible for production if ever it should become necessary. The Aden had been successfully developed by the Design Department at Enfield from the German Mauser MG 213 and manufactured for use in aircraft in the 1950s.[13]

The RSAF in June 1960 took over responsibility for the design of small-arms and related ammunition along with the staff and facilities of the Armament Research and Development Establishment. In consequence about sixty non-industrial staff were added to the section at Enfield. Fazakerley was now about to close and all its outstanding orders were passed to the RSAF together with plant and equipment including a Boma automatic woodworking machine for the production of rifle butts. This replaced 17 former machines with the consequent saving of labour and floor space. Substantial redundancies were now found necessary at Enfield and working hours were reduced from 44 to 42. An unofficial stoppage in protest followed. New techniques continued to be introduced, including that year the cold forging of small-arms barrels and work on automatic barrel production. A change was made to oil-fired boilers. The number employed at Enfield had fallen by 31 March 1961 to 1,741.[14]

The work formerly undertaken at Fazakerley was soon absorbed by the Enfield factory which was now engaged on the production of the buffered tripod and General Purpose Machine Gun. Orders for small-arms had become so limited that the established production method, using semi-skilled labour working with machine setters, had been largely superseded by the employment of all skilled operatives.

The Design and Development Department had been engaged both on weapons and ammunition, on various machine-guns and secondary armament mainly for the Chieftain tank. The department had produced a kit rapidly to convert the .303 inch No 4 rifle to fire the NATO 7.62mm high-velocity rimless ammunition.[15]

Enfield was among the group of 'men's factories' and, although women were employed on certain ancillary tasks, the production work was normally restricted to men. Like most of the ROFs, the factory was finding difficulty in the recruitment of skilled labour. Many of those employed had spent a lifetime in the factory, like Leslie Crouchman of Beaconsfield Road who had joined the Inspectorate of Armaments at the age of fourteen in 1911. Another long-service man was George Wheeler who found work at the factory in 1938 after his employer, a nurseryman, had transferred his business to Somerset. Yet another veteran was George Huttlestone of Ponders End, the third generation of his family to work at the RSAF. He had become a 'Lockie' in 1914, then he had volunteered for the Navy, but had finally returned to the factory in 1936.[16]

Recruitment was still largely through the apprenticeship system. The factory held an annual exhibition of apprentice work and a prize-giving ceremony, which was attended by the parents and representatives of the colleges where the apprentices were taking part-time day release courses. Apprenticeships were granted to boys under 17 mostly to be trained in light engineering. The best of those selected had the prospect of transfer to a student grade of apprenticeship and ultimate qualification as a technician or technologist. Thus the factory knew the value of recruiting its skilled labour in this traditional manner. But the factory had a further means of filtering its skilled recruits. After serving their time the apprentices were discharged, said Stan Carroll. They had to find work at other firms and to work there for at least a year to prove they knew their trades. Only then could they apply at the RSAF. During the Second World War things were different; the apprentice gained his indentures after five years (as against seven before the war) and was taken on at the factory straight away. Eric Kent told how the engineering apprentices were moved around the factory to learn 'the job from A to Z'. Their training was 'second to none' and RSAF apprentices, applying for jobs elsewhere, were always given preference. One former apprentice, S. Thorpe, had won a scholarship to the Enfield Technical Institute in 1919 and had served his apprenticeship from 1921-26. He joined the Armaments Design Department after working first in the production shops, then in the drawing office and subsequently with the Chief Inspector of Small-Arms. Thus his valuable skills had been slowly accumulated within the works. Even in May 1969 the Director of the Inspectorate of Armaments could declare, at the RSAF, that there were excellent opportunities in government departments for engineering training and post-apprenticeship employment.[17] Writing of the apprenticeship scheme many years after his retirement, Bert Hunt, formerly secretary at the RSAF, claimed that it ranked high in the industry and was well regarded within the ministries of Supply and Defence. 'The investment in time and resources', he said, 'coupled with the contribution from the supervisors, was of the highest order and was supported by the management. The scheme embraced the well-being of the apprentices as well as their skill development and education'.

Conscription ended in 1962 and the contraction predicted by Aubrey Jones in 1957 began at Enfield Lock. By the end of March the number employed had fallen to 1,673. Much of the northern part of the site was closed down and production concentrated in the buildings at the southern end. The factory in the Sixties was equipped with numerically controlled machinery, 'loaded and unloaded (rather than operated) ... by unskilled workers', said Eric Kent; this reduced labour costs and 'increased work tempo'. Stan Carroll viewed the use of numerical control somewhat differently. Enfield was in the forefront in the use of these machines, he pointed out, because the RSAF had benefited from government ownership in that 'you could go and look at other people's organizations and they weren't too commercially worried... but there was nothing inherent in the hardware... that entailed de-skilling the workforce'. There had been many problems of reliability at first, he said, and skilled setters had been employed as machine operators. These machines were expensive but adaptable and could save on manpower, they could produce a small series of pieces or even a single piece and so were useful now that mass production was no longer the principal object. The wages of the production workers still rose by regular annual increases leaving the wages of the maintenance workers persistently trailing behind, giving them a constant sense of grievance. They came out on strike in the summer of 1963 only to have their claim deferred. Some seven months later, in January 1964, they came out on unofficial strike once more, protesting against the seven months delay and complaining that their pay had fallen even further behind that of the production workers. The maintenance workers were yet again on strike in August 1969 after new negotiations on a pay claim had dragged on for a further nine months.[18]

The No 4 Short Magazine Lee Enfield had come at last to the end of its life. In August 1966, with full naval honours, and a piped lament, a coffin containing the rifle was ceremonially lowered into the earth on the rifle range at Bisley. It had now been rendered out of date by the Self-Loading Rifle. We had gone to war in 1939 with the Short Magazine Lee Enfield which was replaced during the war by the No 4 rifle. The Lee Enfield in both its forms had been the result of modifications at Enfield of the Lee Metford rifle. The rapid bolt action of the Lee Enfield and its ten-round magazine had put the German k98 Mauser, with its awkward bolt action and five-round magazine, at a disadvantage. Though the Mauser had a good nine-inch bayonet as against the Lee Enfield's unwieldy, 1914-18, sword-bayonet or the 'ridiculous prong that went with the No 4 rifle', bayonets were not much used, except enthusiastically or dramatically in practice, in the Second World War. ROF Woolwich ceased production in September 1966.[19]

The devaluation under the Wilson government in November 1967 was followed by cuts in expenditure on the armed forces. Britain's 'East of Suez' role was abandoned and orders for expensive defence weaponry were cancelled. Tighter controls now had to be put on arms procurement. At this juncture a committee, chaired by Sir John Mallibar, recommended that the financial structure of the ROFs be changed to render them more like commercial undertakings. Vote financing was to be replaced by a trading fund and each factory was to control its own capital assets. Yet the ROFs were not free to design and develop products which, in the judgement of the local management, might find a profitable market. They could only respond to the demand for products, perhaps designed elsewhere, which had to be manufactured to very precise specifications. Major new orders and orders for items of military or political importance, from clients other than the British government, had to be referred to ROF headquarters. The function of the ROFs remained as before; to provide the main reserve capacity to meet possible emergencies. Each factory director was responsible for planning and controlling his own production processes and was empowered to authorize purchase of plant up to £5,000. He remained according to Bert Hunt, 'wholly responsible, together with his team, for all equipment purchased and commissioned, albeit after gaining approval...on the larger purchases. It was to this end', added Bert Hunt, 'that Derek Hatton-Williams led a team from the factory to look at the latest manufacturing techniques throughout the United States small arms industry in the early Sixties'. What was seen influenced future design and manufacturing systems. A factory director could recruit industrial workers and non-industrial staff in basic grades, declare redundancies up to 25 and discharge unestablished men on disciplinary grounds. He was empowered to negotiate piece-work prices but not basic wages and salaries.[20]

Further considerable reorganization was carried out at Enfield between the years 1967-9 when all the workshops east of the canal were transferred to the main part of the factory west of the canal.

Staff was divided between industrial and non-industrial Civil Servants, but only a small proportion of the industrial labour force was permanent and pensionable. Advertisements for production workers on piece-work held out 'prospects of pensionable employment after a qualifying period'. Training was offered, canteen facilities and sick pay. The same facilities were offered to skilled tool and gauge examiners, their pay was £19 7s for a forty-hour week in November 1968, but it was going up all the time; it reached £20 11s by the following May and £22 2s by January 1970. The wage of canteen assistants in 1968 was £9 9s for a forty-hour week. Even the Army Department Constabulary, where preference was given to ex-servicemen, offered only *prospects* of pensionable employment. Millwrights, maintenance fitters and electricians and other skilled workers were offered the

102 The main entrance to the factory in the late 1960s. The building to the left inside the gates was the search box, the prefabricated buildings behind housed the Central Personnel and Records Office

same conditions; their pay, by early in the year 1971, was £23 1s for a forty-hour week. Pensions for established workers were payable at sixty, but even appointments in managerial posts were initially unestablished. Unestablished staff were entitled to gratuities on leaving, after five years service. Piece-work prices were

103 Numbers 1, 2 and 3 Officers' Quarters faced Lock Lane, now Ordnance Road. In number 1, in 1881, lived Capt William McClintock the assistant superintendent with his wife and mother and son Robert aged 7, Emily Harknet, a cook and general servant, aged 19 and Annie Ellis, a housemaid, aged 16. Number 2 Officers' Quarters, in 1881, housed Capt Hunter with his wife, two children, a forty-five-year-old female cousin and his thirty-nine-year-old sister-in-law with her son. He employed a cook, a housemaid, a nurse, and as gardener a 44-year-old RSAF pensioner who lived separately in a house in the garden. The senior clerk John Black lived in number 3

negotiated at factory level. The trade unions, the Amalgamated Union of Engineering and Foundry Workers, the TGWU and the GMWU had their own Whitley negotiating machinery. Typical of the old type of shop steward was Stanley Clear (TGWU) chairman of the Enfield Lock branch, living in a nice thirties-type house in Monastery Gardens and employed as an examiner. He had been a regular soldier and a regimental boxing champion. Some element of the old conservatism survived in the factory, men such as Edwin Farrow who had joined the Inspection Department in 1935 after eight years in the RAF. He was chairman of Enfield Highway Conservative Club from 1955 to 1969.

The RSAF remained busy working a three-shift system in May 1970 and constantly advertised in the local press for machine setters and production workers. A

number of machines made in Switzerland were in use. Arthur Sapsworth was presented, in November 1970, with a gold watch by the Société Genevoise d'Instruments de Physique, for having operated their machines at Enfield over the longest recorded time. By the beginning of the year 1971 the Army Inspection Department had been renamed the Army Quality Assurance Department. In August that year the scientific staff, together with those of the Explosives Research and Development at Waltham Abbey, were protesting in typical trade union language, against a 'recent insulting pay offer'.[22]

There was a proposal in July 1971 to demolish the former 'colonels' houses' at 130-4 Ordnance Road. No 134, once used to house the superintendent, had in recent years been converted to an apprentice hostel, No 132 had of late years been used as an apprentice training school. The whole site was sold to the Borough Council for £300,000 in August 1972. The Council, having purchased, allowed the properties slowly to deteriorate and by March the following year they had been wrecked and desecrated by vandals. Thus another link with the proud past of Enfield Lock was allowed to sink ignominiously into ruin. The last of the three properties, number 130, was not demolished until May 1975. At that time the Council was negotiating to buy, from the MoD, a number of smaller properties at the rear of the site. The Ministry of Defence even proposed the demolition of the houses in Government Row to make way for new quarters for factory staff. Enfield Lock Ratepayers and Residents Association stood firm for their preservation and with strong support from Enfield Preservation Society the area, including the lock, the lock-keeper's house and cottage, together with Government Row, was designated a conservation area in November 1976.[23]

The demand for skilled workers remained high. Advertisements appeared in the local press for setters to set capstans, milling, drilling and grinding machines, and electricians for the installation, maintenance, repair and servicing of machinery, also tool and gauge examiners, all at a basic rate of £27.95 (June 1972) plus shift allowance. Production workers were also sought for machining process and assembly work, for which a period of initial training was offered. Labourers were sought at £19.55 for a 40-hour week in January 1972, their basic pay within twelve months had increased by no less than £2.60, over 13 per cent. Craft apprenticeships were offered for boys under 17 years old.[24]

One-day strikes were called in July 1972, to take place every Thursday, and an overtime ban was imposed by some eight hundred industrial Civil Servants in

104 (opposite) The surgery can be seen behind the reception and personnel department. It was equipped with everything that might be needed for accidents or illness at work. It was staffed by two full-time nursing sisters and two part-time. A local GP visited daily to check people off sick or returned from sick leave. Workers on the ranges were periodically examined for lead poisoning. The surgery had its own ambulance, seen here standing forlorn long after the closure. There were two drivers, both worked in the factory and were trained in first aid

105 Sports day — Barry Rudge, Gordon Bird and others

support of an increase of pay. The men had refused an increase of 7.5 per cent, John Mackie, the Labour member for Enfield East, strongly supported their claim. Another dispute was brewing at the same time, men in the service department had refused to accept a government bonus scheme and in consequence were losing from £6 to £10 a week. The men appealed for help to various MPs, but it was Jeremy Thorpe who stood forward as their champion. He arranged for his local Liberals, including Dr Eric Ridge, chairman of Enfield North Liberals and Ian Stuart, chairman of the Association of Liberal Trade Unionists, to meet the shop stewards to work out a solution. Questions were asked in the House of Commons. The following month, one of the shop stewards, returning to work after a day's sick leave (perhaps exhausted from working 70 hours the previous week) was told by the foreman that in order to ensure his complete recovery he would not for the time being be allowed to do any overtime. Immediately a special meeting was called and an overtime ban was imposed. A month later the shop steward (now perhaps sufficiently rested) was given permission to resume overtime, but by this time the management had eliminated the need for overtime by the imposition of a shift system, which left the men still losing money.[25]

The factory was busy by October 1973 and advertising for both full-time and part-time workers in many departments. Part-timers were offered the choice of morning, afternoon or evening work. There were also vacancies for boys leaving school to join the factory training scheme. Wage rates had been increased again from July that year. It was a time of raging inflation. An attempt by the Government to control wages and prices broke down in February 1974, the NUM once again came out on strike and many factories were reduced to working a three-day week. Others like the RSAF fared better, having their own generators. That year came the first of many local rumours that the factory could be closed within a few years. It was of course categorically denied by an MoD spokesman. The following month the AEU brought its members in all the local factories out on a one-day strike to protest against a £75,000 fine imposed on the union for its contempt of the National Industrial Relations Court.[26]

As I have said, many who worked at the factory at this time are reluctant to create an impression that industrial relations there were worse than average. Something of a family atmosphere persisted. The factory had a drama group, camera club and shooting club. Dances old and modern were held, socials and Christmas parties for the children and the retired. There was also a railway club and a fishing club, both still running years after the closure. A sports day was held annually, swimming in the basin before it was filled in and swimming races in the canal in front of the boiler house and a slippery pole race. The sports field was at the back of the factory and was used for cricket and football, according to Charlie Brown, 'it was a nice family afternoon out'. The library remained in use and employees had allotments on a site behind the factory. The traditional 'Lockie' still survived, one such was Frank Turner who had followed both his father and his grandfather into the factory. His father had been employed as a carpenter, his grandfather as a french polisher. Frank was born near the factory, in Warwick Road, and had been christened at the font in the factory church (demolished 1928). He went to school at Chesterfield, then to Ponders End Technical School, from where he was selected as an engineering apprentice when he was fourteen. During the Second World War he worked on Brens which were assembled at Enfield from bodies made in Canada and barrels made in Australia. The gauges to ensure the accuracy of these parts were provided by craftsmen like Frank Turner. Many around Enfield Lock were so concerned in 1975 about that old font, seen lying neglected outside the church in Ordnance Road, that they campaigned for its return to the factory where it was housed outside the directors' dining room.[27]

106 The factory church interior showing the font

The weapons being produced at Enfield in 1973/4 included the 30mm Rarden, the 7.62mm FN(L1A1) rifle, the 30mm Aden and the 7.62mm General Purpose Machine Gun. The Rarden was designed by Norman Brint of the Design Department at the factory. It was a unique concept at the time. It was purposed for use in the Royal Armoured Corps, Fox and Scimitar vehicles and also on the FV 432 armoured personnel carrier. But in 1971, five years after the contract had been given to the RSAF Enfield, it still had not been perfected, no one apparently had appreciated the difficulties involved. Even in the autumn of 1972, when some of the first production models had failed their tests, the designers remained confident that the faults were minor and continued production. It was not stopped until November 1973 when further failures disclosed that some of the gauges were inaccurate because of mistakes in the design and drawing, also there were factory errors in some of the parts.

Production had no sooner been restarted in 1974-5 than the Army made the unsettling discovery that its FV 432, when equipped with its smart new gun and turret, could not squeeze in all the ten soldiers normally carried in the back. In consequence one role for the gun had to be abandoned and several hundred were cancelled from the Army's order.

This was not all, by March 1975 the delivery rate for the eagerly awaited Rarden was only half the expected number. The result was that many of the Fox and Scimitar armoured vehicles had to be put into store because there were no guns to put on them.

The story, said the *Times* (25 June 1976) had a happy ending and the Ministry declared itself entirely satisfied. It was observed however that while the complexity of developing the new weapon was fully appreciated, 'we must record our serious concern about the air of undue optimism and the apparent lack of forward planning and firm control throughout its history'. Nevertheless the Rarden remains in use in the late Nineties. Another startling new weapon, the SA 80, developed at Enfield by the Design Department over four years at a cost of £2m, was introduced to the Army in June 1976. It was hoped it would become the standard rifle for NATO in the 1980s. It was lighter, at 9lb 1oz, shorter, at 30 inches, and cheaper, £160 against £260, than the Self Loading Rifle (SLR) which had been in service in Britain for 18 years. It also had the advantage that it required only one kind of bullet for all its functions as against three required with the SLR. Army marksmen also declared it to be more accurate than the SLR, which troops in Northern Ireland had found awkward to handle in difficult circumstances. This new Individual Weapon was intended also to replace the Sterling sub-machine gun (SMG) and, with a bipod, longer barrel and other modifications it would convert into the Light Support Weapon (LSW), to replace the Bren. NATO proposed to begin a two-year series of trials in 1977. This Enfield Weapon System (EWS) was a family of weapons designed to give the British soldier his first ever personal weapon, the SA80, capable of automatic fire. There was also a training 'look alike' which fired only single shots. The same ammunition (0.280 inch calibre) was fired in each of the weapons. After extensive trials by NATO, alongside weapons from other countries, NATO decided to adopt 5.56mm calibre ammunition and the EWS, redesigned to use this, was ordered by the Army to meet its needs for the 1980s. Enfield could claim at this time that it was fully equipped with modern machine-tools, many of them numerically controlled. There was a precision casting foundry, a press shop and heat treatment and metallic plating plants.

The year 1977/8 saw a continuous growth of sales by the eleven ROFs, particularly to overseas customers, and in recognition the company received the Queen's Award for Export Achievement for the second time. Commercial and overseas work now accounted for more than half ROF sales. A capital investment programme was formulated based on an increased volume of sales to £370m a year. There was a serious setback early in the year 1979, owing to the cancellation, following the overthrow of the Shah, in 1978, of the contract to supply 1,200 battle tanks to Iran. This badly hit the factories at Leeds, Enfield Lock and Nottingham. A further setback was caused by the long Civil Servants' strike in the factories.[28]

Notes to Chapter 8

1 Putnam and Weinbren p97
2 *ibid* p99
3 Beith, J. H *The Story of the ROFs 1938-48* HMSO 1949
4 SUPP 5 1068
5 *Times* 22 Oct 1947
6 *Times* 15 Au 1949
7 SUPP 5. 1068, 1069
8 SUPP 5. 1069
9 SUPP 5. 1070
10 SUPP 5. 1071
11 SUPP 5. 1068, *Scotsman* 4 Oct 1957, WO 286.23
12 *Times* 26 Mr, 5 Je 1957, SUPP 5. 1071
13 SUPP 5. 1087, information Stan Carroll, Putnam and Weinbren *op cit* p105
14 SUPP 5. 1068
15 SUPP 5. 1089
16 SUPP 5. 1089, *Gazette* 26 Mr 1963, 21 F 1964
17 *Gazette* 12 Ap 1963, 3 F 1967, 6 Ja 1966, 23 My 1969
18 *Gazette* 17 Ja 1964, 22 Au 1969, Putnam and Weinbren *op cit* 108
18 *Gazette* 5 Au 1966, PP 1967/8 xxxv, *Times* 28 Ja 1976
20 Cmnd 4713
21 eg *Gazette* 21 F, 5 S, 3 Oct 1969, 8 My, 23 Oct 1970
22 *Gazette* 15 Ja, 6 Au 1971
23 *Gazette* 16 Jl 1971, 11 Au 1972, 9 Mr 1973, 15 My 1975
24 *Gazette* 12, 26 Ja 1973
25 *Gazette* 10 N, 1, 15 D 1972
26 *Gazette* 5 Oct 1973, 11 Ja, 5 S 1974, 9 N 1973
27 *Gazette* 5 Ja, 17 Jl 1975
28 *Times* 15, 25 Je 1976, PP 1977/8.688, *Times* 11 N 1982, PP 1980/1.103, *Times* 11 N 1982, information Stan Carroll

Chapter 9

Privatization & Closure

The Government in 1980 set up a study group to look into the operations of the ROFs and the suspicions of Labour MPs were immediately aroused. It was suspected that the intention was to sell off parts to private enterprise, that the Government was seeking to reward private industry. A government spokesman in reply asserted that the study was being undertaken to see whether greater freedom could be found to enable the Ordnance factories to make commercial arrangements with the private sector. But rumours continued to circulate and were enthusiastically spread in the local press with lurid stories of the intended sale of Enfield Lock to foreigners, especially to Germans.[1]

The first Thatcher government had shown a widespread interest in the privatization of all nationalized industries in order to induce greater competition and to increase efficiency.

The predicted changes were finally announced by John Nott, as Secretary of State for Defence, in the Commons in May 1982 and they brought immediate and inevitably hostile, trade union reaction. The Government was to retain control over the eleven factories (now employing 19,400 workers) but the involvement of private capital was to be invited and the whole establishment was to be transferred out of the Civil Service. Profits had fallen in the year 1981/2 to only £1m, from £15m in the previous year and £40m in 1977.[2]

Frederick Clarke, until recently at IBM, was appointed in November to head this new state military equipment supplier. Each factory would continue for the time being to be managed under a trading fund until legislation could be introduced to run the group under the Companies Acts. Any plans the Government might have had to sell the RO factories were deferred until after the general election, for there was no mention of the necessary legislation in the Queen's speech. Consultations indeed had only just begun with the formidable trade unions over the intention to bring the factories under the Companies Acts.[3]

Profitability revived in the year 1982/3 to £68m, this on total sales of £449m of which 43 per cent were exports. *The Times* was of the opinion that the growing export potential of the company would interest the City. Lazard Bros, merchant bankers, were asked to investigate and recommend whether the factories could be privatized. It was widely assumed, said *The Times,* that they could command a price of £300m if sold to the public.[4]

107 The square in front of the clock tower where the ceremony for the launch of the Enfield Weapons System took place

The second reading of the Ordnance Factories and Military Services Bill was before the Commons in January 1984. It provided that Civil Servants employed in the ROFs would transfer to the new organization on the same terms and conditions and that there would be no loss of pension rights. A holding company would be established with four subsidiary companies, one of which would produce small-arms and small-arms ammunition. The Rocket Motor Executive at the south site Waltham Abbey would come under the rocket motor subsidiary which also controlled Westcott and Summerfield. The Bill was challenged with a twenty-four-hour strike by many of the 19,000 workers at the 11 ROFs, to protest against privatization. More than 500 strikers lobbied MPs as the Commons debated the second reading. Royal assent was received in November.[5]

By 1984, with the MoD about to spend millions on the new SA80 and its ammunition, for deliveries were due to begin in June, it was reported in the *Times* that a German company, Heckler and Koch, had perfected what was described as a superior weapon employing a revolutionary technology capable of firing rounds without cases and at a greatly increased rate of firing. It was further said to be ‘ the

108 Many dignitaries were invited for the launch of the Enfield Weapons System in October 1985. The Duke of Kent is seen here talking to Stan Carroll. A large marquee was erected in the square by the clock tower, seen in the illustration opposite

greatest breakthrough since the abandonment of gunpowder'. Advocates believed that it could be brought into mass production within two years and the firm was said to be in secret negotiation with the French and American governments. It looked like bad news for Enfield, for the British government was expected to spend £2,000m on the SA80 over the ensuing twenty years. It was hoped that plans were too far advanced to change.[6]

The German government had entered its weapon (the G11 made by Heckler and Koch) in the 1977 NATO trials. Although the system embodied several revolutionary features (caseless ammunition, a recoiling barrel, and auto three-round fire) at that time it had not been developed sufficiently and was withdrawn from the trials. The German Army had been supplied with its standard G3 rifle some years later than the British Army had received its L1A1 (FN) rifle and its re-equipment was therefore less urgent. Heckler and Koch were confident by 1984, that they were close to solving their problems with the G11 and tried to persuade agencies in the UK that the British Army re-equipment with EWS should be replaced by a later re-equipment with the G11. Their attempt failed. It is interesting that the RSAF had co-operated for many years with Heckler and Koch to the limit possible, for the RSAF was not at the time an entirely commercial company. However Royal Ordnance PLC, shortly after gaining commercial status, purchased Heckler and Koch.

The new Enfield Weapon System (EWS) was formally handed over to be used by the Ministry of Defence as the standard infantry weapon for the British Army at a ceremony in October 1985. A large marquee was put up in the square by the clock tower, recalled Barbara Bassett. Many dignitaries attended including the Duke of Kent and the work-force assembled around the square to witness the proceedings. Tanks and personnel carriers were brought in to enliven the event. The EWS thus was able to continue a tradition which had existed since 1854 that Royal Ordnance should supply the principal infantry weapon for the British Army. The new weapon was produced at Enfield, it was the first new rifle for twenty years, but the market was very competitive for this class of weapon and buyers had plenty of choice. Enfield was also working on other important weapons: the Mauser Cannon fitted to the Tornado fighter, the spotting rifle for the light Anti-Armour Weapon (LAW 80), the Multi-Purpose Weapon (ARWEN), in which it was hoped to interest law enforcement agencies throughout the world, the Hughes Chain Gun (machine-gun) and the 25 mm. Aden to be used in the Harrier GR5 jump jet. That same year a new £5m metal finishing plant came into operation at Enfield; it was claimed to be 'among the most advanced metal finishing facilities in Western Europe'.

The SA80 was to continue to have a troubled life. The Commons Defence Committee as late as 1993 described it as badly designed, it was said still to suffer 32 faults which had put soldiers' lives at risk during the Gulf War of 1991. By the time this report was published more than 332,000 had been supplied to MoD at a cost of £384m, modifications it was predicted would cost a further £24m. The Commons Defence Committee demanded that the manufacturer and not the taxpayer should be saddled with the bill.[7]

Government plans to withdraw the 240-strong MoD police force that had long guarded the RSAF and to replace it with a private security firm, as part of the privatization plan, had to be quickly abandoned after criticism from the All-Party House of Commons Defence Committee. Unlike the MoD police, the private security firms had no access to arms during any emergency. 'We were appalled by the present levels of security at RSAF Enfield', said the committee report. Security at Enfield 'suggested neglect and complacency' said the MPs after a visit to the factory on 17 May 1984. The night-time security they found consisted of one man on duty, and he was responsible both for answering the telephone and watching the premises. The factory in places was protected by a single wire fence and rear windows were exposed. The MPs however were relieved to be told that financial

109 The plan to replace MoD police by a security firm in 1984 proved to be a fiasco. Its night-time security turned out to be one man armed with a telephone. £500,000 had to be spent on a new police station and gates, seen here abandoned in 1990

approval had now been given for physical security improvements costing £500,000 and that further measures were intended.[8] A new police lodge and security gates were constructed.

The eleven ROFs were expected to be sold by a stock market flotation early in the year 1986. They remained dependent on the MoD for 60 per cent of their orders, despite this more than 90 per cent of British defence orders were placed in the private sector. It was a situation perhaps too vulnerable to attract buyers. In addition a poor trading performance in the year 1984/5 had forced Royal Ordnance to spend £15m out of its reserves and this had created fears that the flotation would be at a price which failed to reflect public investment in the company. The House of Commons Defence Committee warned the Government that its desire for a quick sale should not override the importance of getting the best value for money. The *Times* thought that it was increasingly likely that the flotation of RO, planned for June, would have to be delayed. City analysts, moreover, were far from convinced that the company was ready for privatization, but the Government feared that, should the RO miss its target date, it would be lost in a flood of privatizations in the autumn. Circumstances intervened however, that summer the market flotation had to be cancelled at the last moment following a public dispute over a £100m Challenger tank order. The MoD had planned to

110 Only two years before the sell-off a large sum of money was expended to raise the standard of the canteen. This photograph of the refurbished canteen was taken after the factory was closed

award the contract to RO Leeds, avoiding public tender, but Vickers objected, demanding to be allowed to bid. The Leeds factory was then sold to Vickers complete with the Challenger order against the wishes of RO.[9]

At Enfield however an order for 175,000 SA80 weapons in June 1985 seemed to promise enough work to keep the men busy for the next four years. Yet by October 1986 there were widespread rumours, denied by the management, that the RSAF was to close. The management asserted that the factory was expecting a second order for the SA80 which would ensure work there up to 1992. Meanwhile they were busily recruiting both skilled and semi-skilled workers. Tim Eggar, Member for Enfield North, declared that the factory had the longest order book in his constituency. An ROF spokesman, more realistically, pointed out that the MoD's requirement for conventional arms was declining and that the future must lie with orders from abroad.[10]

By the beginning of January 1987 the idea of flotation had been abandoned, the Government had decided to sell RO to a single buyer and a series of factory visits began immediately, at which potential bidders were given a detailed review and breakdown of the company. Their main worry concerned who would bear the costs of future reconstruction and redundancies. The *Times* was of the opinion that if purchasers were forced to accept these liabilities the final bids could fall below £100m. As if to emphasise this opinion, a week later the *Times* referred to RO as the £100m valued company which, in view of the later and much inflated valuations bandied about, poses some questions about the financial judgement of that newspaper. The unions were concerned that there should be no job cuts. At the time 1,200 were employed at Enfield Lock and 320 at Waltham Abbey. The transport workers spokesman described morale at Enfield as 'very, very low'.

Four of the country's biggest firms had expressed an interest; they were GNK, BAe, Trafalgar House and Ferranti, the bidders had until the end of February to confirm an interest of purchase and until the end of March to submit a contract of sale. The unions were worried that Trafalgar House might outbid its rivals, for that company had no direct interest in the manufacture of arms and it was feared that it might axe jobs and sell off sites. Industrial action to protect jobs was however thought improbable, at least in the near future. Martin Upham, prospective Labour candidate in Enfield North, pledged that a future socialist government would re-nationalize RO.[11]

With the appointment of Roger Pinnington as new chief executive at the beginning of the year, the work-force at the Land Weapons Division at Enfield came increasingly under pressure. Pinnington had threatened to lock the gates and transfer production elsewhere if work on the contract to supply 175,000 SA80s did not speed up. To date (Feb 1987) only half had been delivered and work had not started on a £3m order for armoured personnel carrier weapons and 25mm. cannon for the new Harrier aircraft. A story had appeared in a national newspaper that the work-force at Enfield was failing to meet important deadlines and that a warning had been given that the factory could miss out on multi-millon-pound contracts unless there were huge improvements in the work rate. A decision on the matter appeared to be only a month away, a transfer of production to another site after the sale of RO was entirely possible. 'Either under private ownership or as a state-owned PLC', said a local newspaper, 'the future of the site at Enfield is precariously in balance'. Martin Upham had written to the Defence Minister, George Younger, expressing the fears of the workers about the track record of some of the prospective buyers and asking for guarantees about future working conditions.[12]

It was nevertheless announced early in March 1987 that RO had been awarded a £100m contract for a second tranche of the order for the SA80 short rifle/machine-gun for the British forces, although that order could not be confirmed until after privatization. There had been four firms unsuccessful in securing the contract. The *Times* thought that the Government was likely to be criticized for using the order to fatten-up RO in preparation for privatization, but it was of the opinion that the order would help to secure both the future of the Royal Small Arms Factory at Enfield, and the success of privatization. Closure costs if incurred at Enfield could amount to £30m.

Despite winning this second order, job cuts at the 'loss-making Enfield plant' were announced: 140 redundancies would be necessary as part of changes in working practices and manpower levels. A company statement however said: 'suitable redeployment of skills within the site will significantly reduce the number of those whose services will be terminated'. It was hoped to avoid compulsory redundancies.[13]

Meanwhile the privatization quickly threatened to degenerate into a fiasco. By the end of the first week in March Trafalgar House had lost interest and dropped out. Why this company should lose interest so soon, if there was truth in forthcoming allegations that many millions were to be made simply from the disposal of the two north London sites, is difficult to understand. Another of the bidders had submitted a zero offer. This was probably Ferranti which pulled out by the middle of the same month. The Government had always been reluctant to hand RO over to the Manchester based group, believing that the merger would make it too heavily dependent upon government defence orders. According to the *Sunday Times* neither of the remaining bidders looked likely to offer anything like the £150-200m which had been hoped for.[14]

Roger Pinnington was pressing the Government to insist on a price related to the asset sheet value of £235m, but insiders thought the Government would accept a far lower sum, one even suggesting a price as low as £50m. A spokesman for Rothschild, the merchant bank handling the sale, denied that this was a possibility. 'If the bids are next to nothing (the merchant banker too, it seems, knew little of the potential value of the sites) the Government simply will not sell', he told the *Sunday Times* and suggested that the bidders were trying to talk the price down. The bidders alleged fears concerning the ability of RO to survive when exposed to competitive tendering.[15]

The privatization however was completed with the sale to British Aerospace (BAe) on 2 April 1987, GKN had offered about £160m, BAe £190m. BAe was now free to sell any part of RO, though not to foreign groups, and to close down any product lines. It believed that it could improve overseas sales and develop high technology production. BAe warned that RO's existing operations would require additional investment and rationalization. The *Times* thought that both the Government and BAe had reason to be pleased about the sale of RO, for the Government had got a better price than it might have expected while BAe had acquired a business which well complemented its own. At the same time there was a feeling, shared by Roger Pinnington, that the Government had taken an easy way out, that RO was fit enough to stay independent, and that the flotation plans could have been resurrected at a later date.[16]

The workers at Enfield now feared impending closure and although union leaders

111 'The gaunt sprawling buildings', it was said, 'were the wrong size, shape and design for the modern armaments business'

were told that BAe had no immediate closure plans the company had already begun a three-month study to identify action that might be needed to put Royal Ordnance into shape. By early May, BAe was considering whether it should continue with the £100m second contract for the SA80. The future of the RSAF at Enfield Lock looked likely to depend upon the outcome. The contract had been accepted at a fixed price, following competitive tendering, although the factory had failed to overcome technical and production difficulties with the earlier contract for 175,000 rifles. Deliveries of these had fallen badly behind schedule and the MoD was contracted to pay only the lower price agreed for the second contract for each rifle now supplied under the first contract. This looked likely to jeopardize the profitability of the first contract. BAe and RO had until the last week in June to decide whether to go ahead with the second rifle contract and had not ruled out the possibility of building a new production line, not necessarily at Enfield, in order to produce the rifle more economically using automated equipment.[17]

The closure of the RSAF was thus precipitated by the decision of RO, as part of BAe, to take up the £100m MoD second contract. Closure was announced on 12 August 1987. It was to take place towards the end of the year 1988. Closure costs, it was estimated, would be £13.5m. Enfield workers would be considered for employment at other RO sites. A more modern plant was to be constructed within the existing Ordnance factory at Nottingham where £12m was to be invested. The manufacture of small-arms would then be transferred there. The Institute of Professional Civil Servants blamed the closure on the years of neglect at Enfield Lock by the previous owners, the MoD. The 'gaunt sprawling building', it was said, 'was the wrong size, shape and design for the modern armaments business'. Dr Maurice Dixson, chief executive at RO since it had been taken over by BAe, dismissed with some difficulty suggestions that a crucial factor in the decision to close Enfield Lock was the potential value of its 110-acre site, nevertheless it was potentially valuable for it lay 'in an area of mixed residential and commercial development, close to the M25'.[18]

BAe asserted that its decision to close resulted from a determination to become an internationally competitive manufacturer of rifles and machine-guns and to do that the RSAF, which was a relatively old and high cost unit, had to go. This was essential because the Ministry of Defence had now adopted a policy of competitive tendering and was in any case buying less equipment. Only 20 per cent of the previous year's RO production had been exported and Dr Dixson thought that RO should become sufficiently competitive to secure overseas markets, for it was dangerous to rely too exclusively on MoD purchases. The world's small-arms industry had become fiercely competitive and saturated with supplies, Third World countries were setting up their own small-arms industries and many producers were short of orders. At the moment, declared Dr Dixson, 'we are not competitive in small gun making'. The new facility at Nottingham would be designed to manufacture other types of small calibre weapons whereas the Enfield plant had been geared solely to the SA80. This had 140 components of which 70 were made in-house. The Nottingham operation would concentrate on 12 key components including the barrel, while all other items would be purchased from elsewhere, if necessary from foreign suppliers.[19]

A mass meeting was held in the factory that August with tough talking from angry union leaders. They called for support from the other Royal Ordnance factories, urging resistance to the shut-down, but as one worker put it after the meeting: 'If we do take action they might close the gates earlier. It's over and it's not the workers' fault'.

Several hundred people gathered outside the Civic Centre to be present for a special meeting of Enfield Council called by the Labour minority in September to formulate practical policies to help the 1,200 people due to lose their jobs. A motion was finally agreed which said;

> This Council notes with deep concern the recently announced intention of British Aerospace to close the RSAF. In view of the fact that this factory is known world-wide and is part of Enfield history and that closure will mean the loss of 1,200 jobs, this council opposes the proposals... In the event of the closure being finally confirmed Enfield Council would wish to see manufacturing opportunities kept alive on the site ....[20]

The unions continued their campaign for increased severance pay. A mass meeting towards the end of November was held in the social club in Government Row and addressed by Jack Dromey who was leading the unions' joint negotiating team. He considered the £750 over and above redundancy pay, plus pay in place of notice offered, to be *derisory,* especially in view of the fact that British Aerospace was likely to make 'tens of millions of pounds' by turning the factory site into 'a yuppie paradise'. Another mass meeting in December decided upon acceptance, by a show of hands, of new terms offered but about a third of those present opposed the deal. Co-operation in reaching production targets, and flexible working within reason, was agreed. Each worker below retirement age would now receive civil service redundancy pay plus around £4,000. The unions continued to press BAe for a financial justification of the closure, threatening court action if it was not forthcoming. Concern was expressed that workers were left in uncertainty about when they would be laid off, for they were unable to leave without forfeiture of severance pay.[21]

The impending closure at Enfield Lock caused considerable anxiety concerning the future of the Pattern Room. It was the world's most comprehensive collection of firearms, and was considered of vital importance to weapon designers and academics. Housed in a two-storey Victorian building within the factory were some six thousand weapons ranging from a William III flintlock musket to the SA80, and all in working order. Herbert Woodend, the curator told the *Daily Telegraph* that it contained the best collection of Warsaw Pact firearms outside Moscow. The Pattern Room derived its name from the ancient system of arms production in which a perfectly correct example of every model was selected and sealed as the ultimate pattern for subsequent manufacture. The collection is now housed at Nottingham.[22]

The Public Accounts Committee in their 48th report 1987/8 reviewed the MoD arrangements for the sale on the basis of a report by the National Audit Office in November 1987. The committee recognized that the sale achieved the Government's objective of transferring the company to the private sector within the prescribed time-scale, but noted the possibility that the Royal Ordnance sites at Waltham Abbey and Enfield might be redeveloped. It expressed concern that British Aerospace could make a substantial profit on the sale or redevelopment of these sites.[23]

After the Government's announcement in 1982 that the ROFs were to be privatized the City estimated that the price would be more than £300m, although in June 1986 Rothschild expected only £100m from the flotation. Prior to the sale to BAe, Royal Ordnance had announced its decision to close the Waltham Abbey site and was preparing planning applications for that and the Enfield North site which abuts it. Subsequently RO declared an intention to close the Enfield factory. The open market values of the Waltham Abbey and Enfield sites, before decontamination and clearance, had been estimated in March 1984 at £2m and £1.5m. These estimates assumed that their location within the Green Belt would inhibit the redevelopment of the sites.[24]

In preparation for the original sale two valuations had been conducted, one by RO and one by the independent surveyors, Weatherall, Green and Smith. This firm was asked to look at the redevelopment potential. Its report was cautious, pointing out that much of the land was in the Green Belt and that planning permission could take several years. A revised figure of £30-£40m was suggested for both Waltham Abbey and a 35-acre section of the Enfield site; these sites were then either vacant or scheduled for closure.

This figure was not passed on to the four companies, Sir Clive Whitmore, the Permanent Under- Secretary, arguing that it was a matter which should be 'left to the judgement of the companies'. The Minister was also of the opinion that more might be obtained from the companies by allowing them to do their own valuations.

There was some sense in this for Sir Raymond Lygo, BAe's chief executive had then agreed to pay £190m, nearly double what had been expected.

In spite of the opinion of the independent surveyors however, Royal Ordnance directors had decided, early in 1986, that if planning permission could be obtained, the 35 acres at Enfield and another 200 acres at Waltham Abbey were worth, not just £30m-40m, but two of three times that amount.

Ivan le Gallois, a senior planner with the Epping Forest District Council, which was then responsible for planning consent for most of the land, received an invitation in April 1986 to visit the sites. He and his colleagues were entertained to lunch in the company's dining rooms, by RO top officials, and afterwards were given 'an extremely professional presentation' about the redevelopment potential of the sites. He remembered seeing page after page of high quality diagrams and maps showing that out of the 235 acres, the company planned light industry and housing on 153 acres, while the remainder were to be given to the local authorities as a park.

That planning application was turned down but, after a public inquiry early in 1988, Epping Forest District Council agreed that 89 acres could be redeveloped. The British Aerospace decision to close the rest of the Enfield site would make a further 110 acres available but, even without this extra land, BAe, it was thought, would recoup most of the £190m it paid for RO by redevelopment of the site.[25]

A report issued by the All-Party Public Accounts Committee in October 1988 was sharply critical of the sale to BAe, accusing the Government of breaking a promise made to the committee that the taxpayer would get the best possible return for the sale of Royal Ordnance. It alleged that the Ministry of Defence was forced to sell the company too quickly in order to meet the Government's privatization timetable. The report was a serious embarrassment to the Secretary of State, George Younger, who had defended the sale on the grounds that it was the best possible deal the Government could get. The criticisms were contained in a draft report prepared by Robert Sheldon (Labour), the chairman of the committee. BAe was accused of asset stripping and planning to sell the land occupied by the two factories for £450m. When it bought RO the property and land was valued in published accounts at less than £5m. One Conservative member of the committee was quoted as saying, 'We are now finding that BAe apparently has made a much better judgement than MoD'. The Treasury rejected these criticisms, insisting that it had obtained an alternative use valuation in addition to the open market value based upon the existing use, but this, which took account of decontamination costs, was less than the existing valuation and was therefore not provided to the bidders. Any assessment of the amount which might be realized, the Government said, must at present be purely conjectural having to take into account how much and what type of redevelopment might be permitted as well as site preparation costs and RO restructuring costs.[26]

The machinery remaining at the factory was sold by auction in a well-attended three-day sale in November 1988. Included were over eight hundred modern toolroom and production machine-tools, sheetmetal working and plastic injection moulding machinery, foundry and heat treatment plant; in all over 3,700 lots. No overall celebration marked the final closure, just small groups of employees meeting to mark a sad farewell to each other and to the old factory. For years the site remained deserted, nothing flourishing but the weeds and incipient trees ever ready to take advantage of man's neglect.

The announcement on 30 May 1989 that BAe and Trafalgar House had formed a joint venture company, Lee Valley Developments, to redevelop the two sites created another political storm since it gave rise to the possibility of profits from development worth twenty times the 1984 valuation. The Labour Party described the business as 'sleazy'. The joint venture company proposed to seek planning permission for a mixed commercial and residential development. The £450m valuation, quoted by Sheldon, had been made in August 1988 by S.G. Warburg the securities company, and although a National Audit Office report does say that Warburg considerably overestimated the value of the sites, it found that, with outline planning permission, the sites would have been worth £63m on the open market in 1989. The report says that the MoD should have carried out a revaluation of the sites. It urged that, in future, where land is considered to have development potential, serious consideration should be given to clawback provisions in order to protect the taxpayers' interests.[27]

Plans for redevelopment of the area put at risk a vital part of Enfield's heritage. Enfield Preservation Society urged that the best buildings should be listed to ensure their survival and early in 1989 machine shop No 1 and other buildings around the former pool were given Grade II listing. The Borough Council nevertheless rejected a request from the Preservation Society and English Heritage that the site should be included in the Enfield Lock Conservation Area. The developers had put forward a plan which would leave only half of the listed machine shop, intended then to house a museum, to recall to future generations the long and glorious history of the factory.

Local elections in 1994 returned a Labour administration which at once rescinded the plan, demanding more social housing. Deadlock ensued and continued for more than a year. Boundary changes which came into force on 1 April 1995 brought the whole of the site within the jurisdiction of the L.B. Enfield. In April 1996 Trafalgar House sold the site to the local developers, Fairview Estates and the Council set up a joint panel under the chairmanship of Councillor John Jackson to work out a new scheme.

The latest application (September 1996), from Fairview Homes demands even more housing on the site but envisages the re-opening of the canal arm and basin and the conversion of more, though too few, of the mid-nineteeth century buildings.

Notes to Chapter 9

1 *Times* 18 D 1980, eg *W Herald* 3 S, 16 Jl 1981
2 *Times* 21 My 1982
3 *Times* 5 N, 11 N 1982
4 *Times* 23 Je, 17 Oct 1983
5 *Times* 17 Ja 1984
6 *Times* 15 Ja 1984
7 *Gazette* 19 Ap 1984, *RO Financial Report 1985*, 'Into the Front Line', *Daily Mail* 25 Je 1993
8 *Daily Telegraph* 27 Jl 1984, *Times* 15 Ja 1984
9 *Times* 5 Oct 1984, 23 My, 19 S 1985, 6 Mr, 3 Ap 1987
10 *Gazette* 19 Ap 1984, 13 Je 1985, 23 Oct 1986
11 *Advertiser* 15 Ja 1987
12 *Independent* 4 F 1987, *Advertiser* 5 F 1987

13 *Times* 6 Mr 1987
14 *Times* 16 Mr 1987
15 *Sunday Times* 8 Mr 1987
16 *Times* 2, 3, 11, Ap 1987
17 *Gazette* 9 Ap 1987, *Financial Times* 7 My 1987
18 *Times* 14 Au 1987
19 *Financial Times* 14 Au 1987
20 *Gazette* 3 D 1987
21 *Gazette* 3, 24 D 1987
22 *Daily Telegraph* 9 N 1987
23 H/C 1987/8 206, 162
24 National Audit Office *MoD Sale of Royal Ordnance plc* 1987
25 *Sunday Times* 23 Oct 1988
26 *Sunday Times* 23 Oct 1988, *Times* 19 Ja 1989
27 *Financial Times* 31 My 1989, 3 My 1990

Index

Numbers in **bold** refer to pages with illustations